Roza Serova
Galiya Rakhimova
Eugenia Stasilovich

Standardization and metrology in the production of building materials

Roza Serova
Galiya Rakhimova
Eugenia Stasilovich

Standardization and metrology in the production of building materials

ScienciaScripts

This book is a translation from the original published under ISBN 978-620-2-09612-6.

Publisher:
Sciencia Scripts
is a trademark of
Dodo Books Indian Ocean Ltd. and OmniScriptum S.R.L publishing group

120 High Road, East Finchley, London, N2 9ED, United Kingdom
Str. Armeneasca 28/1, office 1, Chisinau MD-2012, Republic of Moldova, Europe
Printed at: see last page
ISBN: 978-620-8-01853-5

Table of Contents

Introduction

The problem of quality of products, works and services is relevant for all countries regardless of the maturity of their market economy. To become a participant of the world economy and international economic relations, it is necessary to improve the national economy taking into account global achievements and trends.

The modern stage of development of the world community is characterized by a progressive movement towards the creation of a common market of goods and services, international division of labor, collective solution of safety and environmental protection issues. Such movement is impossible without ensuring the uniformity of technical documentation requirements for component parts, manufacturing methods and equipment without ensuring the uniformity of measurements, etc.

An important direction of activity of standardization organizations - work on the creation of normative documents and practical activities on certification of products and production. These measures are designed to ensure the interests of consumers, environmental protection and safety.

One of the most important characteristics of products is their quality. Quality is a set of characteristics of an object that relates to its ability to meet established and anticipated needs. An object can be an activity or process, a product, an organization, a system or an individual, as well as any combination of them.

Standardized methods and measuring equipment are used to objectively assess the quality of products manufactured by different manufacturers. The availability of a standardized product testing scheme allows obtaining objective criteria for assessing the quality of products and stimulates manufacturers in a market economy to improve the quality of products. The desire of the producer of products to realize them in the shortest possible time stimulates him to continuously improve the quality of products.

As defined by ISO: Standardization is an activity aimed at achieving an optimum degree of order in a given field by establishing provisions for universal and repeated use with respect to real and potential problems.

Each physical quantity can be measured by different methods. In a particular situation the obtained result differs in accuracy. Analysis of the results allowed to choose the method that allows to obtain the most accurate result under the given conditions at minimum cost. This method is approved as a standard.

Thus, standardization is subject to objects, processes or solutions that are repeated in the form of variants (or have such a possibility), standardization is not subject to non-repeated phenomena or repeated but without variants.

The most important result of the introduction of standardization is an increase in the degree of conformity of standardization objects to their functional purpose.

1 Fundamentals of standardization

1.1 General characteristic of standardization

Standardization is an activity aimed at the development and establishment of requirements, norms, rules, characteristics, both mandatory and recommended, ensuring the consumer's right to purchase goods of proper quality at an acceptable price, as well as safety and comfort of work.

The purpose of standardization is to achieve an optimum degree of order in a particular field by establishing provisions for universal and repeated use with respect to actual or potential tasks.

The main objectives of standardization:

- protecting the interests of the consumer and the state in matters of quality and range of products, services, processes, their safety for human life and health, and environmental protection;

- improving product quality, compatibility and interchangeability;

- Elimination of technical barriers to production and trade,

ensuring product competitiveness;

- saving of human and material resources, improvement of economic indicators of production;

- improving the country's defense capabilities;

- ensuring the safety of state facilities, taking into account the possibility of natural and man-made disasters and other emergencies;

- harmonization with international, regional and national standardization systems of other countries.

Specific objectives of standardization relate to a particular area of activity, industry of goods and services, a particular type of product, enterprise, etc. [1].

Objectives of standardization:

- ensuring mutual understanding between developers, manufacturers, sellers and consumers (customers);

- establishment of optimal requirements to the nomenclature and quality of products in the interests of the consumer and the state, and ensuring the safety of the environment, life and health of people, preservation of their property;

- development of product compatibility and interchangeability requirements;

- establishment of metrological norms, rules, regulations and requirements;

- creation and introduction of technical and economic information classification systems;

- implementation of the RK legislation by methods and means of standardization.

Standard (from English norm, sample), in the broad sense of the word - a sample, standard, model, taken as a reference for comparison with them other similar objects.

The standards define the procedure and methods for planning quality improvement at all stages of the life cycle.

Standard is a normative and technical document that establishes the basic requirements for the quality of products, the rules of their development, production and application.

As defined by ISO/IEC 2: A *standard* is a document, developed by consensus and approved by a recognized body, that establishes, for universal and repeated use, rules, general principles relating to different activities or their results, which aims to achieve an optimum degree of regularization in a given field.

Quality standards must be agreed upon by each enterprise when concluding contracts for the supply of products to the market, participating in tenders, competitions, obtaining state orders and preferential lending. According to the legislation, the development of state standards is entrusted to the leading enterprise in the industry. The developed project receives feedback, is corrected, approved and becomes mandatory for all existing entities [2].

When developing standards, take into account the already adopted international and regional standards and ensure compliance of the requirements of standards with the norms of legislation, optimality of requirements included in the standards.

Standards are formulated clearly and explicitly to ensure that their requirements are unambiguously understood.

Objects of standardization are: products, processes (works), services, having the prospect of repeated reproduction and (or) use. *Standardization area* is a set of interrelated objects of standardization. For example, the area of standardization is the industry of building materials, the objects of standardization in it - different types of materials.

Standardization takes place at different levels. The *level of standardization* differs depending on which geographical, economic or political region of the world adopts the standard.

International standardization is standardization that is open to the relevant authorities of any country.

Regional standardization is an activity open to the appropriate authorities of the States of one geographical, political or economic region of the world.

National standardization - standardization in one particular state. In this case, national standardization can be carried out at different levels: state, sectoral, in a particular sector of the economy (for example, at the level of ministries), at the level of associations, enterprises, institutions.

Management of works on standardization in the Republic of Kazakhstan shall be carried out by the authorized body on standardization, metrology and certification.

Authorized body on standardization, metrology and certification:

1) forms and implements a unified state policy in the field of standardization;

2) coordinates the activities of public administration bodies, individuals and legal entities in this sphere;

3) participates in works on interstate, international, regional standardization;

4) organizes and carries out state supervision of compliance with mandatory requirements of normative documents on standardization;

5) establishes a state system of standardization;

6) organizes professional training and retraining of personnel in the field of standardization;

7) establishes the procedure for the application of international, regional, national standards, rules and recommendations on standardization;

8) interacts with individuals and legal entities, technical committees on standardization;

9) establishes in the state standards, rules and recommendations of the state system of standardization general organizational and methodological and general technical rules for carrying out works on standardization, forms and methods of interaction of individuals and legal entities with each other, with state authorities;

10) provides translation of normative documents on standardization into the state and Russian languages and their confirmation;

11) establishes the procedure for examination of normative documents on standardization for compliance with the requirements of standards, international and regional norms.

Bodies of state administration of the Republic of Kazakhstan shall participate in the works on standardization within their competence.

Physical and legal persons shall organize and carry out activities on standardization in accordance with this Law, other normative legal acts regulating relations in the field of standardization within their competence, and may create for the performance of these works appropriate subdivisions and services for standardization, including technical committees on standardization.

The authorized body on standardization, metrology and certification represents the Republic of Kazakhstan in international and regional organizations on standardization, metrology, certification, quality management and accreditation within its competence.

The organizational structure of the state standardization system consists of:

1)   from the authorized body on standardization, metrology and certification, its territorial subdivisions and subordinate enterprises;

2)   bodies of state administration of the Republic of Kazakhstan within their competence in the field of standardization;

3)   individuals and legal entities, including technical committees, experts - auditors on standardization;

4)   of the State Standards Fund of the Republic of Kazakhstan.

Regulatory documents on standardization, operating in the state standardization system of the Republic of Kazakhstan, include:

1)   State Standards of the Republic of Kazakhstan ST RK (hereinafter - State Standards);

2)   state classifiers of technical and economic

information - GK TEI (hereinafter referred to as state classifiers of technical and economic information);

3)   interstate standards - GOST (hereinafter interstate standards), classifiers of technical and economic information, rules and recommendations;

4)   international, regional and national standards, classifiers of technical and economic information, technical conditions, rules, instructions, regulations, guidelines, methodological instructions and recommendations on standardization applied in accordance with the established procedure;

5)   standards of scientific and technical, engineering societies and other public associations;

6)   recommendations;

7)   firm standards;

8)	specifications;

9)	industry standards.

Requirements established by normative documents on standardization shall be based on the achievements of science, engineering and technology and shall not contradict the legislation of the Republic of Kazakhstan, applicable technical regulations, requirements of international, regional and national standards of foreign countries, rules and recommendations for standardization, taking into account the conditions of use of products, performance of processes (works) and services, working conditions and modes.

When developing normative documents on standardization, international technical regulations and standards or their drafts in the stage of completion are used, except when they do not meet the requirements of safety for life and health of people, environmental protection and technical norms adopted in the Republic of Kazakhstan [3].

Normative documents on standardization for products, processes (works), services subject to mandatory certification shall contain requirements for which mandatory certification is carried out, methods of control for compliance with these requirements, rules for marking and packaging of products, information on certification.

Normative documents on standardization should not be a technical obstacle in production and trade with other states.

Indicators of target purpose and technical compatibility of specific groups and types of products are established during the development and putting the products into production in accordance with the norms and requirements adopted in the territory of the republic, unless otherwise provided by the agreement (contract) for their delivery.

Registration of normative documents on standardization shall be canceled in the prescribed manner, if their norms and requirements are not applied in the territory of the republic, as well as in case of changes in the rules of international trade.

State standards and classifiers of technical and economic information, as well as normative documents on standardization, approved by public administration bodies in accordance with their competence, are not the object of copyright.

Interstate normative documents on standardization shall be developed, coordinated and adopted in the order established by the Interstate Council for Standardization, Metrology and Certification.

1.2 Normative documents on standardization

Normative Document (ND) - a document that establishes rules, general principles or characteristics concerning various activities or their results.

The following types of regulatory documents are distinguished.

Standard - a normative document developed by consensus, approved by a recognized body, aimed at achieving an optimal degree of order in a particular field. A standard establishes for universal and repeated use general principles, rules, characteristics relating to various activities or their results.

Standards can be *international, regional, national, administrative-territorial*. They are adopted by the relevant authorities and used by the relevant range of consumers. Standards are periodically reviewed for changes to keep their requirements in line with scientific and technical progress.

A specification document establishes technical requirements, for a product, service, process. This document usually specifies the methods or procedures to be used to verify compliance with the requirements of this normative document in such situations where it is necessary.

Code of Practice may be an independent standard or an independent document, as well as part of the standard. Code of Practice is usually developed for the processes of design, installation of equipment and structures, maintenance or operation of objects, structures, products.

All of the above normative documents are *recommendatory*. In contrast, regulations are mandatory. A *regulation* is a document that contains mandatory legal norms. A regulation is adopted by a standardization body, as in the case of other normative documents. A type of regulations - technical regulations - a document containing technical requirements for the object of standardization. They can be presented directly in this document or by reference to another normative document.

The ISO Guide presents the following possible types of standards.

A fundamental standard is a normative document that provides general or guiding statements for a particular field. Usually used either as a standard or as a guidance document from which other standards may be developed.

A *terminology standard* in which the object of standardization is terms.

The *standard for test methods* establishes methods, rules, procedures for various tests and related activities (e.g., sampling or specimen collection).

A product standard contains product requirements that ensure that the product is fit for its intended purpose. It may be complete or incomplete. A complete standard establishes not only the above requirements, but also the rules of sampling, testing, packaging, labeling, storage, etc.

Standard for process, standard for service are normative documents, in which the object of standardization are respectively process (e.g., technological), service (e.g., repair work, domestic service).

The *compatibility standard* establishes requirements relating to the compatibility of the product as a whole, as well as its individual parts (parts, assemblies).

State standards are developed for products, works and services, the needs for which are cross-sectoral in nature. They contain both mandatory to the object of standardization requirements, and recommended. To mandatory include: safety of the product, service, process for human health, environment, property, as well as industrial safety and sanitary standards; technical and information compatibility and interchangeability of products; unity of control methods and unity of labeling.

Industry standards (OST) are developed in relation to the products of a particular industry. Their requirements should not contradict the mandatory requirements of state standards, as well as rules and safety standards established for the industry. The objects of industry standardization may be products, processes and services used in the industry, etc.

Enterprise standards (STP) are developed and applied by the enterprise itself. The objects of standardization are usually the components of the organization and management of the enterprise, the improvement of which is the main goal of standardization at this level.

Standards of Public Associations (STO) - normative documents developed, as a rule, for fundamentally new types of products, processes or services; new test methods, etc.

Technical Specifications (TS) are developed by enterprises when it is inexpedient to create a standard. The object of TU can be products of one-time delivery, produced in small batches.

Types of standards:

1)   *fundamental standards* are developed to promote mutual understanding, technical unity and interconnectedness of activities in different fields of science, technology and production;

2)   *standards for products (services)* establish requirements either for a specific type of product or for a group of homogeneous products:

3)   *standards for works (processes)* establish requirements for specific types of works that are carried out at different stages of the product life cycle: development, production, operation, storage, transportation, repair, disposal;

4)   *standards for control methods (tests, measurements, analysis).*

1.3   Functions performed by standardization

In the conditions of market relations standardization performs three functions: economic, social and communicative.

The *economic function* allows stakeholders to receive reliable information about the product in a clear and convenient form. Reference to the standard in the contract terms obliges the supplier to deliver products of quality guaranteed by the standard.

Standardization of test methods allows for comparable product characteristics, which is important in assessing product competitiveness.

The *social function of* standardization is the requirement to include in standards such quality indicators of the object of standardization that promote occupational safety and health and sanitation.

Communicative function is related to the achievement of mutual understanding in society through the exchange of information. This requires standardized terms, interpretations of concepts, symbols, uniform rules of office work[4].

1.4 Principles and methods of standardization

Standardization as a science and as an activity is based on certain initial provisions - principles.

The *principles of standardization* reflect the basic regularities of the process of standards development and their implementation. Seven important principles of standardization can be distinguished:

1) balance of interests of the parties developing, manufacturing, providing and consuming products (services); all parties interested in the standardization of any object should find consensus, i.e. the absence of objections on essential issues from the majority of stakeholders;

2) systematicity and comprehensiveness, i.e. consideration of each object as a part of a more complex system;

3) dynamism and advanced development. The first is ensured by periodic review of standards, amendments, revision, abolition of obsolete ones; the second is ensured by taking into account in the development of standards the development trend and advanced standards of other countries or firms;

4) efficiency of standardization. The application of normative documents (ND) should produce economic, technical, informational or social effect;

5) Prioritize the development of standards that promote safety, compatibility and

interchangeability of products and services;

6) harmonization interstate and international so as not to create barriers to international trade;

7) clarity of the wording of the provisions of the standards to avoid ambiguity in the wording of the ND.

The method of standardization is a set of techniques by which the goals of standardization are achieved.

Basic methods of standardization:

1) *Arrangement of standardization objects* - reduction of diversity. Arrangement of standardization objects is a universal method in the field of standardization of products, processes and services. Arrangement as diversity management is primarily associated with the reduction of this diversity. It includes:

a) *systematization* - scientifically-based classification, clustering and ranking of aggregates of objects. Systematization consists in arrangement in a certain order and sequence, convenient for use. The simplest form of systematization is the arrangement of systematized material in alphabetical order (in reference books, bibliographies, etc.). In technology is widely used numerical systematization by number order or in chronological sequence. For example, in a standard, in addition to the number, enter digits indicating the year of its approval.

б) *selection* - reasonable selection of objects that are recognized as feasible for further production and application;

в) *symplification* - reasonable selection of objects that are recognized as inappropriate for further production and application;

г) *Typization* - creation of *typical* objects that can be subjected to some non-radical transformations or adaptations for specific applications. Typization is a method of standardization, which consists in the establishment of standard objects for a given set, used as a basis (base) in the creation of other objects close in functional purpose. Typization is developed in three main directions: standardization of standard

14

technological processes; standardization of standard designs of general-purpose products; creation of normative and technical documents establishing the procedure for carrying out any work, calculations, tests, etc. Typing of technological processes is the development and establishment of a technological process for the production of similar parts or assembly of similar components or products of one or another classification group. Typing of technological processes is caused by the need to reduce the unreasonably large number of them on the same-type parts or assemblies. Very often the technological process is developed anew without taking into account the existing experience. When changing the object of production the whole volume of technological developments is repeated anew and a significant part of technological processes duplicates previously developed processes. Typing of technological processes during their optimization allows to exclude these disadvantages and accelerate the process of production preparation.

д) *optimization* - finding the optimal indicators of standardization objects according to some criteria (for example, economic efficiency) with the help of special two-dimensional or multidimensional mathematical optimization models;

2) *Parametric standardization.* Product parameter is a quantitative characteristic of its properties. Parametric standardization is to select and justify the appropriate nomenclature and numerical value of parameters;

3) *Unification of products.* Product unification is understood as the activity on rational reduction of the number of types of parts and units of the same functional purpose. It is based on classification and ranking, selection and symplification, typification and optimization of elements of finished products. The unification of rows of parts, assemblies, units, machines and devices is based on their constructive similarity, which is determined by the commonality of the working process, working conditions of products, i.e. the commonality of operational requirements.

The following types of unification are distinguished: standard-size, intra-size and inter-type.

Type-size unification is used in products of the same functional purpose, differing from each other by the numerical value of the main parameter.

Intra-type unification is carried out in products of the same functional purpose, having the same numerical value of the main parameter, but differing in the design of component parts.

Inter-type unification is carried out in products of different types and different design (for example, unification of longitudinal milling, planing, grinding machines among themselves).

Unification works can be carried out at the following levels: factory, industry, interindustry and international.

The level of unification of products or their components is determined by means of a system of indicators, of which the coefficient of applicability at the size level, calculated as a percentage, is mandatory:

$$K_{np}^{m} = \frac{n - n_{o}}{n} \cdot 100\%,$$

where n is the total number of product sizes;

n_0 - number of original sizes.

The application of unification allows to noticeably reduce the volume of design work and reduce the design time; to reduce the time for preparation of production and mastering of new products; to increase the volume of production due to specialization, as well as the quality of products.

However, unification accompanied by certain costs requires economic justification. Unjustified unification can have a negative effect, in particular, when it is necessary to use the nearest large unified parts, causing unjustified by operational conditions increase in weight, dimensions and labor intensity of machine manufacturing.

To optimize standardization means to standardize such designs and their size ranges, at which the total efficiency in the sphere of production and operation would be the greatest.

The main directions of unification are:

- development of parametric and size ranges of products, machines, equipment, devices, units and parts;

- development of standard products in order to create unified groups of homogeneous products;

- development of unified technological processes, including technological processes for specialized production of products of inter-industry application;

- limiting to a reasonable minimum the range of products and materials authorized for use.

The results of work on unification are formalized in different ways: they can be albums of standard (unified) designs of parts, units, assemblies; standards of types, parameters and dimensions, designs, marks, etc.

4) *Aggregation* is a method of creating machines, devices and equipment from separate standard unified units, repeatedly used in the creation of various products on the basis of geometric and functional interchangeability. Aggregation provides expansion of the area of application of machines, devices, equipment of different functional purpose by means of their assembly from separate units manufactured at specialized enterprises. These units should have full interchangeability in all operational indicators and connection dimensions. Aggregation makes it possible to reduce the amount of design and engineering work, reduce the time of preparation and mastering of production, reduce the labor intensity of manufacturing products and reduce the cost of repair operations. Aggregate machine tools consisting of unified elements are widely spread. When changing the object of production, they can be easily disassembled and new machines for machining other parts can be assembled from the same units.

5) *Complex standardization.* At complex standardization the development of a system of interrelated requirements for both the object of complex standardization as a whole, and to its main elements in order to optimally solve a particular problem is carried out.

6) *Advance standardization.* This method involves the establishment of increased in relation to the already achieved in practice level of norms and requirements for objects of standardization, which according to forecasts will be optimal in the future.

1.5 International and regional standardization

International standardization is a set of international standardization organizations and products of their activities - standards, recommendations, technical reports and other scientific and technical products. There are three such organizations: International Organization for Standardization - ISO (ISO), International Electrotechnical Commission - IEC (IEC), International Telecommunication Union - ITU (ITU).

The International Organization for Standardization is the largest and most authoritative of the above-mentioned organizations. Its main purpose is stated in the ISO Constitution: "...to promote the development of standardization on a world scale to ensure international exchange of goods and mutual assistance and to increase cooperation in the fields of intellectual, scientific, technical and economic activities".

The main purpose of international standards is to create at the international level a unified methodological basis for the development of new and improvement of existing quality systems and their certification.

In recent years, ISO has devoted much attention to the standardization of quality assurance systems. The practical result of efforts in these directions is the development and publication of international standards. When developing them, ISO takes into account the expectations of all stakeholders - producers of products (services), consumers, government, scientific, technical and public organizations.

ISO's strategy in recent years has emphasized trade and economic activities that require the development of appropriate solutions in the interest of the market and an operational model that makes full use of the potential of information technology and communication systems, taking into account, first and foremost, the interests of developing countries and the emergence of a global market on equal terms.

ISO standards, accumulating the advanced scientific and technical experience of many countries, are aimed at ensuring the unity of requirements for products that are the subject of international commodity exchange, including the interchangeability of component parts, uniform methods of testing and quality assessment of products.

The users of ISO International Standards are industry and business, government and non-governmental organizations, consumers and society at large.

ISO International Standards do not have the status of being mandatory for all member countries. Any country in the world has the right to apply or not to apply them. The decision on the application of ISO International Standard is mainly related to the degree of participation of the country in the international division of labor and the state of its foreign trade. In the standardization system of the RK about half of ISO International Standards have found application.

ISO standards differ in their content in that only about 20% of them include product-specific requirements. The bulk of the normative documents deal with safety requirements   ,

interchangeability, technical compatibility, product test methods, and other general and methodological issues. Thus, the use of most ISO International Standards assumes that specific technical requirements for a product are established in contractual relationships.

ISO and IEC jointly develop ISO/IEC guides that address various aspects of conformity assessment activities. The voluntary criteria contained in these guides are the result of an international consensus on best practices and approaches. Their application promotes continuity and orderliness in conformity assessment worldwide and thereby facilitates international trade.

In this way, the principle of "One standard, one test, recognized everywhere" is put into practice.

Although international standards are developed on the basis of consensus and voluntary recognition of their requirements, in practice compliance with them is essentially mandatory, as it is a criterion of competitiveness and admission to the international market.

International standards have become an effective means of eliminating technical barriers to international trade, as they have acquired the status of documents defining the scientific and technical level and quality of products.

Over the last five years, the level of use of international standards has increased from 15% to 35%, and in such industries as machine building, metallurgy, transportation and communications, it has exceeded 40%.

1.5.1 ISO's forward-looking objectives

ISO has defined its objectives by highlighting the most relevant strategic areas of work:

- establishing closer links between the organization's activities and the market, which, above all, should be reflected in the choice of priority developments;

- reduction of total and time costs as a result of increased efficiency of administrative staff, better use of human resources, optimization of workflow, development of information technologies and telecommunications;

- Effective assistance to the World Trade Organization through the introduction of a programme focusing on the gradual conversion of product specifications into ISO standards;

- stimulating the "self-sustaining" elements of the above program: encouraging the creation of new standards for industry, developing relations with the WTO on the terms of providing the necessary technical assistance. In particular, it is envisaged to promote in every possible way the inclusion of requirements for supplied products by states in ISO international standards, which should have a positive impact on the recognition of conformity assessment;

- Concern for improving the quality of national standardization activities in developing countries, where the focus is on leveling up standardization levels.

In the future, ISO plans to expand the scope of its technical services. It has identified three priority opportunities: to promote the adoption of widely used industry standards developed outside ISO as international normative instruments; to identify priority standardization needs in specialized areas; and to increase flexibility in planning standards development work in response to changing market and national conditions.

In addition, a fairly rapidly growing area of international standardization continues to be services, where the 9000 series of standards will increasingly be applied.

A number of major governments are transferring responsibility for the development and implementation of standards used for government procurement to the private sector. In this context, ISO is exploring opportunities for international standardization in the non-governmental sector. In the future, cooperation between ISO, IEC and which complements the activities of these organizations and promotes effective standardization programmes in the field of information technology and telecommunications will become increasingly important [5].

1.5.2International standards for environmental management systems ISO 14000

The emergence of ISO 14000, a series of international standards for environmental management systems in enterprises and companies, has been called one of the most significant international environmental initiatives. The ISO 14000 system of standards, unlike many other environmental standards, is not focused on quantitative parameters (volume of emissions, substance concentrations, etc.) and not on technologies (requirement to use or not to use certain technologies, requirement to use "best available technology"). The main subject of ISO 14000 is the environmental management system. Typical provisions of these standards are that certain procedures must be introduced and followed in the organization, certain documents must be prepared, a person responsible for a certain area must be appointed. The main document of the series - ISO 14001 does not contain any "absolute" requirements for the environmental impact of the organization, except that the organization in a special document must declare its aspiration to comply with national standards. This nature of the standards is due, on the one hand, to the fact that ISO 14000, as international standards, should not intrude into the scope of national regulations. On the other hand, the predecessor of ISO are "organizational" approaches to product quality, according to which the key to achieving quality is to build a proper organizational structure and distribution of responsibility for product quality. The ISO 14000 system of standards also used the proven model of international standards for product quality control systems (ISO 9000), in accordance with which more than 70,000 enterprises and companies around the world are currently certified. The first standards in the ISO 14000 series were officially adopted and published at the end of 1996. The system of standards is expected to ensure the reduction of adverse environmental impacts at three levels:

1. Organizational - through improving the environmental "behavior" of corporations.

2. National - through the creation of an essential complement to the national regulatory framework and a component of state environmental policy.

3. International - through improving the terms of international trade.

The documents included in the system can be roughly divided into three main groups: principles for the creation and use of environmental management systems (EMS); environmental monitoring and assessment tools; and product-oriented standards.

The key concept of the ISO 14000 series is the concept of an environmental management system in an organization (enterprise or company). Therefore, the central document of the standard is considered to be ISO 14001 - "Specifications and Guidelines for the Use of Environmental Management Systems". Unlike the other documents, all of its requirements are "auditable" - it is assumed that compliance or non-compliance with them by a particular organization can be established with a high degree of certainty. It is compliance with ISO 14001 that is the subject of formal certification.

The main requirements that ISO 14001 imposes on an organization are as follows:

1. An organization should develop an environmental policy, a specific document of the organization's intentions and principles, which should serve as the basis for the organization's actions and for setting environmental goals and objectives. The

environmental policy should be appropriate to the scale, nature and environmental impacts created by the company's activities, products and services. The environmental policy should, among other things, contain statements of commitment to regulatory compliance, as well as to "continuous improvement" of the environmental management system and to "pollution prevention." The document should be communicated to all employees of the organization and made available to the public.

2. The organization shall establish and follow procedures for determining significant environmental impacts. The organization shall also systematically address all legal requirements related to the environmental aspects of its operations, products and services, as well as requirements of another nature (e.g. industry codes).

3. Taking into account significant environmental impacts, legal and other requirements, the organization should develop environmental goals and objectives. Goals and objectives should be as quantitative as possible. They should be based on environmental policy, and defined for each function (area of activity) and level of the organization. The views of 'stakeholders' should also be taken into account in their formulation.

4. To achieve its objectives, the organization should develop an environmental management program. The program should define responsible, means and timeframes for achieving the goals and objectives.

5. An appropriate responsibility structure should be defined in the organization. Sufficient human, technological and financial resources should be allocated to ensure the operation of the system. There should be a person responsible for the operation of the environmental management system at the level of the organization, whose duty should be to report periodically to the management on the operation of the EMS.

6. A number of staff training requirements must be met, as well as emergency preparedness.

7. The organization should monitor or measure key parameters of those activities that may have a significant impact on the environment. Procedures should be

established to periodically verify compliance with applicable legal and other requirements.

8.   There should be a periodic audit of the environmental management system to determine whether it meets the criteria set by the organization and the requirements of ISO 14001, and whether it is implemented and operating properly. The audit can be conducted either by the company itself or by an external party. The results of the audit are reported to the company's management.

9.   The organization's management should periodically review the operation of the environmental management system in terms of its adequacy and effectiveness. Necessary changes to the environmental policy, objectives and other elements of the EMS should be considered. This should take into account audit results, changed circumstances and the desire for 'continuous improvement'. In general, the requirements of the standard are based on an open cycle of "plan - implement - verify - revise plan".

The standard implies that the environmental management system is integrated with the overall management system of the organization. The standard does not require that persons responsible for EMS work have no other responsibilities, or that documents related to environmental management are allocated to a special document management system [6].

ISO 14000 standards are "voluntary". They do not replace legal requirements, but provide a system for determining how a company affects the environment and how legal requirements are met.

An organization may use ISO 14000 standards for internal purposes, for example as an EMS model or an internal audit format for an environmental management system. It is intended that the establishment of such a system provides an effective tool for an organization to manage the totality of its environmental impacts and bring its operations into compliance with a variety of requirements. Standards can also be used externally to demonstrate to customers and the public that the environmental management system is up-to-date. Finally, an organization can obtain formal certification from a third (independent) party.

As can be inferred from the experience of ISO 9000 standards, the desire to obtain formal registration is likely to be the driving force behind the introduction of environmental management systems compliant with the standard.

International standardization is necessary for the successful implementation of trade, economic and scientific and technical cooperation between different countries. The lack of international standards and the difference of national standards for the same products offered on the world market are barriers to the development of international trade.

The main task of international scientific and technical cooperation in the field of standardization is harmonization, i.e. harmonization of the national standardization system with international, regional and progressive national standardization systems of foreign countries in order to improve the level of RK standards, the quality of domestic products and its competitiveness in the world market.

In 1987, ISO adopted a package of international standards ISO 9000, aimed at a uniform approach to addressing product quality issues in enterprises.

At present, the objects of standardization of this series have been considerably expanded to cover not only elements of quality systems, but also a number of other criteria.

The fundamental standards of this system are:

1)   ISO 9000-1. Standards for quality assurance. Guidelines for the selection and application of specific standards;

2)   ISO 9000-2. General guidelines for the application of ISO 9001, ISO 9002 and ISO 9003, etc.

An urgent task for ISO is to improve the structure of the standards collection. If earlier, in the early 1990s, standards in mechanical engineering (30 %), chemistry (12.5 %) and electronics and information technology accounted for no more than 10.5 %, now the latter are gaining priority.

ISO IS are not mandatory, i.e. each country has the right to apply them in their entirety, in separate sections or not to apply them at all. However, under conditions of market relations and fierce competition, manufacturers are forced to use IS. The level of use of IS in industrialized countries reaches 80% of the entire ISO standards collection.

1.6 Systems of organizational, methodological and general technical standards in construction

System of design documentation for construction - a set of regulatory organizational and methodological documents establishing general technical requirements necessary for the development, accounting, storage and application of design documentation for the construction of objects of various purposes.

The system of normative documents of the Republic of Kazakhstan in construction is created in accordance with the new economic conditions, legislation and management structure on the basis of construction norms, rules and state standards in this area.

The main focus of the newly developed normative documents of the System is to protect the rights and legally protected interests of consumers of construction products, society and the state with the development of independence and initiative of enterprises, organizations and specialists. One of the main means of solving this problem is the transition to new methodological principles, which are increasingly common in the practice of international standardization. In contrast to the traditionally established so-called "descriptive" or "prescriptive" approach, when normative documents provide a detailed description of the design, calculation methods, materials used, etc., newly created building codes and standards should contain, first of all, operational characteristics of building products and structures based on the requirements of the consumer.

Normative documents developed in accordance with building codes and regulations should not prescribe how to design and build, but should establish requirements for construction products that must be met, or goals to be achieved in the design and construction process. The ways of achieving the set goals in the form of volume-planning, structural or technological solutions should be of a recommendatory nature.

The system of normative documents in construction is a set of interrelated documents adopted by the competent executive authorities and construction management, enterprises and organizations for application at all stages of creation and operation of construction products in order to protect the rights and legally protected interests of its consumers, society and the state.

The system, based on the overall objectives of standardization, should contribute to solving the challenges facing the construction industry to ensure that:

- compliance of construction products with their purpose and creation of favorable living conditions for the population;

- safety of construction products for life and health of people in the process of their production and operation;

- protection of construction products and people from unfavorable impacts, taking into account the risk of emergencies;

- reliability and quality of building structures and foundations, engineering equipment systems, buildings and structures;

- meeting environmental requirements, rational use of natural, material, fuel and energy and labor resources;

- mutual understanding in the implementation of all types of construction activities and elimination of technical barriers in international cooperation [7].

The objects of standardization and norming in the system are:

- organizational and methodological and general technical rules and regulations necessary for the development, production and application of construction products;

- objects of urban planning activity and construction products - buildings and structures and their complexes;

- industrial products used in construction - construction products and materials, engineering equipment, equipment for construction organizations and construction industry enterprises;

- economic standards required to determine the efficiency of investments, construction costs, material and labor costs.

The legal basis for standardization and norms in construction is the legislation of the Republic of Kazakhstan, which determines the relationship of participants of investment activities, their rights, duties and responsibility for the quality of products and services. Construction norms, rules and standards are one of the means of intersectoral regulation and management in design and construction in order to implement the requirements of the legislation.

Development of normative documents in construction is envisaged to be carried out on the principles adopted by the state standardization system of the Republic of Kazakhstan and international standardization organizations while ensuring the necessary harmonization and comparability with international standards in the field of construction, construction legislation and standards of technically developed foreign countries.

1.6.1 Objects of standardization in construction

Standardization in construction should be carried out in accordance with the requirements of the State Standardization System.

The objects of standardization in construction are: construction products, industrial products, as well as rules that ensure the development, production and application of these products.

For various objects of standardization in construction, depending on their characteristics develop state standards RK, industry standards - OST and technical specifications - TU.

1. *State standards in construction establish*: organizational and methodological and general technical rules that provide in construction management of production processes of development, production and application of products; requirements for groups of homogeneous products and for specific products.

The objects of state standardization in construction are:

general organizational and methodological requirements in construction;

design documentation for construction;

technological rules of design in construction (standard);

nomenclature of quality indicators by groups of homogeneous products in construction;

requirements of modular dimensional coordination in construction;

requirements for ensuring the accuracy of geometric parameters in construction;

general design rules regulated in standards;

requirements for engineering surveys in construction;

labor safety requirements common to construction;

parameters of buildings and structures of civil, industrial and agricultural purposes, as well as the most massive structures of specialized types of construction, requirements to their elements and interface nodes;

typical technological processes of the main types of construction and installation works;

general technical requirements and rules of acceptance of elements of buildings, structures and construction and installation works;

basic requirements for inventory (mobile) buildings and structures for construction;

General methods of quality control in construction;

general requirements for building, construction and engineering blocks and engineering equipment blocks;

structures and products for civil, industrial, agricultural construction and the most mass structures and products for specialized types of construction;

main types of building materials;

main types of engineering equipment of buildings and structures;

tooling for mass construction and installation works;

molds and tooling for manufacturing of reinforced concrete structures and products of mass application;

main types of fasteners for construction.

2. Industry standards in construction establish:

organizational and methodological requirements of sectoral application;

requirements for groups of homogeneous products and specific products of sectoral application not subject to state standardization.

The objects of sectoral standardization, in particular, may be:

organizational and methodological requirements necessary for the production activities of enterprises and organizations subordinate to the Ministry, including occupational safety requirements;

parameters of buildings and structures of specialized types of construction, requirements to their elements and units;

standard technological processes of certain types of construction and installation works specific to the production activities of individual ministries;

technical requirements and rules of acceptance of elements of buildings, structures and construction and installation works for specialized types of construction;

methods of instrumental quality control specific to certain types of construction;

building structures and products of sectoral application;

construction materials and engineering equipment of buildings and structures of sectoral application;

tooling for specialized types of construction works;

tooling for the manufacture of building structures and products.

3. Technical specifications in construction establish requirements for the manufacture, control, acceptance and delivery of building materials, structures and products, as well as other construction products of specific types (grades) in the absence of state, industry standards.

In particular, technical specifications for building structures and products are developed:

on standard structures and products for transportation, energy, water and other special types of construction, working drawings of which are approved by the relevant ministries and for which the development of standards is not envisaged;

new designs and products for the period until standard designs are developed on their basis;

custom-designed structures intended for mass production.

Custom-designed structures produced under one-off orders for completion of a certain construction object are manufactured directly according to working drawings and delivered in accordance with the standards common for the corresponding group of homogeneous structures.

For groups of products in construction develop standards that regulate for this group of products general technical requirements, acceptance rules, methods of control and other general requirements or standards such as "General Technical Conditions", combining these requirements.

The requirements for specific products are established by standards such as "Specifications", "Design and Dimensions" and "Types, Design and Dimensions".

For groups of building structures, homogeneous in terms of functional purpose and commonality of design solution, standards of the type "Types and basic parameters" are developed, which establish the types of structures, their coordinate dimensions and are intended for use in the design and development of standards or specifications for structures of specific types.

For building structures in the standards of the type "Technical Conditions" establishes the nomenclature of grades of structures and requirements that ensure their quality, drawings of general views with basic dimensions, references to working drawings of structures.

Working drawings of typical designs may be included as part of the standard.

Standards for newly developed and revised standard structures should be developed simultaneously with the working drawings of these structures. When developing standards for standard structures of existing series that do not require revision, the necessary adjustments to the working drawings shall be made at the same time.

Technical specifications for building structures are developed together with working drawings of these structures.

Rows of coordinating modular dimensions, as well as functional parameters of buildings, structures and their elements are established in the standards of the "Parameters" type.

Requirements for the quality of elements of buildings and structures, rules of their acceptance and methods of control are established in standards of the type "Technical requirements, rules of acceptance, methods of control".

1.7 Basic Principles of Building Construction Standards Complex

The development of state normative documents and territorial construction norms shall be carried out in accordance with the norms and rules by research, design and other organizations and associations, as well as creative teams with scientific potential and the necessary experience of practical work in the relevant area. Technical committees on standardization and technical standardization in construction (TCS) participate in the development of documents.

The customer for the development of the document may be the organization charged with its adoption or any other interested organization (enterprise)[8].

The development of normative documents is carried out according to the following stages:

Stage 1 is the organization of document development;

Stage 2 - development of the draft document in the first edition;

Stage 3 - preparation of the draft document in the developer's final version and its submission to the customer;

Stage 4 - consideration, acceptance (approval) and registration of the document;

Stage 5 **is the** issuance of the document.

Organization of the document development includes submission of a reasonable application by the prospective developer, coordination of the scope of work and conclusion of contracts for its execution between the customer, the main developer and co-executors.

At the conclusion of the contract, the customer approves a brief terms of reference for the development of a normative document, which specifies the main goals and objectives of the development, stages of work and deadlines for their implementation, organizations - co-executors, as well as organizations to which the document is sent for review and approval.

Approval of the draft document with state supervision bodies and other organizations specified in the technical assignment for its development shall be carried out by the developer prior to submission of the document for approval.

The draft document shall be sent for approval in the final version of the developer. Consideration of the draft shall be carried out within 30 days from the date of its receipt. Approval is formalized by a letter. The record "Approved with comments" is not allowed.

Disagreements arising during the agreement of the draft document shall be formalized by a joint protocol.

The developer shall submit the draft regulatory document in three copies, one of which shall be the first, with a cover letter and the following documentation in one copy:

-	explanatory note to the draft document with justifications, data on the used results of research works and on the results of comparison of the document with international and foreign standards;

-	draft document sent for review (first edition) and the list of organizations to which the draft document was sent;

-    original opinions of the organizations to which the document was sent for review and a summary of the reviews;

-    minutes of meetings of the STS, conciliation meeting or TAB on consideration of the draft normative document;

-    original documents confirming the approval of the project by state supervision authorities and other organizations or a protocol of disagreements (if the approval was required by the terms of reference);

-    proposals to repeal existing documents or draft amendments to them related to the introduction of a new normative document.

When considering the submitted draft normative document to decide on its adoption, it shall be checked for compliance with the requirements of legislation and existing regulations, general methodological principles of norming and standardization, the requirements of these norms and rules, GSS standards and technical specifications. Based on the results of the document review, the developer makes the necessary clarifications to it.

When a normative document is adopted, the date of its enactment is set. At the same time, the documents, replacing which it was developed, are canceled.

Cancellation of normative documents is carried out by the bodies that approved them.

Specifications for building materials, products, structures and other products of industrial enterprises are developed by `organizations-developers` or manufacturers of these products as an integral part of the design, engineering or technological documentation for its manufacture.

1.8 Peculiarities of standardization of construction materials and products

The methodology of standardization in building materials science and technology takes into account the peculiarities of materials performance and includes standardization as constituent elements:

-   material and structural loads;

-   of environmental influences;

-   dimensions of building products.

1.8.1 Standardization of loads

Loads and impacts are categorized into permanent and temporary.

The *permanent loads* and impacts include: the mass of permanent parts of buildings and structures; the mass and pressure of soils - embankments, and backfills, as well as mining pressure; the force effect of prestressing structures.

Temporary loads are divided into long-term, short-term and special *loads*. *Temporary long-term* (technological) loads include: the weight of stationary equipment (concrete mixers, batchers, hoppers with material); the weight of partitions or other parts of the building, the position of which may change during operation; the pressure of gases, liquids and bulk solids, in tanks and pipelines; long-term temperature effects of stationary thermal equipment; loads on the ceilings of storage rooms.

The following loads and impacts are considered to be *short-term*: dynamic loads from moving equipment; loads on slabs from the weight of people and furniture; atmospheric loads (wind, snow, ice, wave, ice, etc.); temperature and humidity climatic impacts causing deformation of materials in structures, etc.

Special temporary loads arise under the influence of seismic effects, sudden disruptions in the technological process associated with equipment failure. They also include loads resulting from subsidence of the base of structures.

The main characteristics of loads are their normative values, which for permanent loads are taken from design data on geometric and structural parameters of building structures and average values of material density.

The normative values of other types of loads are also regulated in the relevant normative documents. Thus, the territory of the country is divided into a number of climatic regions, each of which has its own values of snow loads and wind pressure, etc.

1.8.2 Standardization of environmental impacts

When developing standards, it is necessary to take into account the following types of physical and chemical effects on materials and structures: climatic, characterized by changes in temperature and relative humidity of outdoor air and other factors; effects of aggressive media, causing corrosion of materials and reducing their durability; humidity regime of the room [9].

Climatic and geophysical indicators should be taken into account when developing normative documents for building envelopes, roofing, wall and facing materials. Standards for these materials should contain requirements for frost resistance, water absorption, etc. Increased humidity regime of premises limits the use of porous structure products for the construction of exterior walls without protective vapor barrier on the inner surface of the wall.

1.8.3 Standardization of dimensions of construction products

The methodological basis for the standardization of dimensions in the design and manufacture of building products for the construction of structures is the modular coordination of dimensions in construction. It allows to carry out the necessary unification of dimensions and thus to ensure interchangeability of a limited number of standard sizes of building products.

The idea behind the modular system is that a structure is dissected in length, width, or height by imaginary coordinating

planes, the distance between which is taken to be equal to some module kM, where M is the size of the basic module, and k is the proportionality coefficient. The size of M is assumed to be 100 mm. The ICRS provides for the preferable use of rectangular spatial

coordination system. For assigning the coordinate dimensions of volume-planning and structural elements, building products and equipment, along with the basic module, derived modules are used, which are obtained by multiplying the basic module by integer or fractional coefficients. When multiplying by integer coefficients, enlarged modules are obtained, and when multiplying by fractional coefficients less than one, fractional modules are obtained.

A distinction is also made between the structural (nominal) dimension, i.e. the design dimension of the element, which differs from the coordinate dimension by the amount of normalized gap, joint or overlap. Design dimensions l, b, h *of* building structures, products and elements can be taken more or less than the coordinate dimensions.

2 Basics of metrology

2.1 Purpose and objectives of metrology. Basic concepts

Over the course of world history, man has had to measure various things, weigh products, and tell time. For this purpose, it was necessary to create a whole system of different measurements necessary for calculating volume, weight, length, time, etc. The data of such measurements help to master the quantitative characterization of the world around us. The role of such measurements in the development of civilization is extremely important. Today no branch of national economy could function correctly and productively without the use of its measurement system. After all, it is with the help of these measurements the formation and control of various technological processes, as well as quality control of products. Such measurements are necessary for various needs in the process of development of scientific and technical progress: for accounting of material resources and planning, and for the needs of internal and external trade, and for checking the quality of products, and for increasing the level of labor protection of any working person. Despite the diversity of natural phenomena and products of the material world, for their measurement there is the same diverse system of measurements, based on a very essential point - comparison of the obtained value with another, similar to it, which was once taken as a unit. In this approach a physical quantity is regarded as some number of units accepted for it, or, to put it differently, in this way its value is obtained. There is a science that systematizes and studies such units of measurement - metrology. As a rule, under metrology is meant the science of measurement, the existing means and methods that help to observe the principle of their unity, as well as the ways to achieve the required accuracy.

The rapid development of metrology occurred at the end of XX century. It is inextricably linked with the development of new technologies. Thus, we can say that metrology studies:

1)   methods and means for accounting for products in terms of the following: length, mass, volume, flow and power;

2)   measurements of physical quantities and technical parameters, as well as

properties and composition of substances;

3)   measurements for control and regulation of technological processes.

There are several main directions of metrology:

1)   general theory of measurement;

2)   system of units of physical quantities;

3)   methods and means of measurement;

4)   methods of determining the accuracy of measurements;

5)   the basics of ensuring the uniformity of measurements, as well as the basics of uniformity of measuring instruments;

6)   measurement standards and exemplary measuring instruments;

7)   methods of transferring unit dimensions from sample measuring instruments and from measurement standards to working measuring instruments. An important concept in the science of metrology is the unity of measurements, which means such measurements in which the final data are obtained in legalized units, while the errors of these measurements are obtained with a given probability. The necessity of unity of measurement is caused by the possibility of comparing the results of different measurements, which were made in different areas, in different time intervals, as well as with the use of different methods and means of measurement.

The objects of metrology should also be distinguished:

1)   units of measurement of quantities;

2)   measuring instruments;

3)   methodologies used to perform measurements, etc.

Metrology includes: firstly, general rules, norms and requirements, and secondly, issues that need state regulation and control. And here we are talking about:

1)   about physical quantities, their units, and their measurements;

2)   principles and methods of measurement and measuring instruments;

3) errors of measuring instruments, methods and means of processing of measurement results in order to eliminate errors;

4) ensuring the uniformity of measurements, standards, samples;

5) to the state metrological service;

6) the methodology of the verification schemes;

7) working measuring instruments.

In this connection, the tasks of metrology become: improvement of measurement standards, development of new methods of precise measurements, ensuring unity and necessary accuracy of measurements.

The word *metrology* is formed from two Greek words *metron* (measure) and *logos* (skill) and means - the doctrine of measures.

Metrology in the modern understanding is the science of measurements, methods and means of ensuring their uniformity and ways to achieve the required accuracy.

Unity of measurements is the state of measurements in which their results are expressed in legalized units and errors are known with a given probability.

National economy of any country requires various information on parameters and characteristics of research and measurement objects in science, production, health care, agriculture, transportation, environmental protection and other spheres of human activity. In modern industry the share of labor costs for measurements makes on average 10 % of total labor costs at all stages of creation and operation of products, and in some industries, in particular, electronic, radio engineering, chemical, reaches 50-60 %.

For this purpose it is necessary metrological support, i.e. establishment and application of scientific and organizational bases, technical means, rules and norms necessary to achieve unity and required accuracy of measurements [10].

The main objectives of metrological support are:

- improving product quality, production management efficiency and the level of automation of production processes;

- ensuring reliable accounting and improving the efficiency of material and energy resources utilization;

- Increasing the effectiveness of measures for the prevention, diagnosis and treatment of diseases, standardization and control of working and living conditions, environmental protection, assessment and rational accounting of the use of natural resources; increasing the effectiveness of international scientific, technical, economic and cultural cooperation.

Technical bases of metrological support are;

- system of state standards of units of physical quantities;

- a system of transferring dimensions of units of physical quantities from a standard to all measuring instruments by means of reference measuring instruments and other means of verification;

- the system of development, production and release into circulation of working measuring instruments ensuring determination with required accuracy of characteristics of products, technological processes and other objects;

- system of obligatory state tests of measuring instruments;

- system of standard samples of composition and properties of substances and materials.

Metrology as a science and a field of practical activity emerged in the Middle Ages. However, its origins date back to ancient times, when man began to feel the need for measurements. At first they were reduced to simple counting. Distance was measured in steps, bow shots, days of travel.

Time was counted in days, phases of the moon and seasons. The number of objects - pieces, dozens, packs, bales.

In order to maintain a given technological process mode, to assess the quality of products, it is necessary to have accurate quantitative information. It can be obtained only with the help of measurements.

In construction it is necessary to measure a wide variety of quantities: linear-angular, mechanical, physical-chemical, thermal, acoustic, optical, etc. To measure these quantities, the construction industry must be equipped with standardized methods and measuring instruments.

Metrology deals with the theory and practice of measurement.

Unity of measurements implies representation of measurement results in legalized units, with measurement errors known with a given accuracy. Ensuring the uniformity of measurements in the country, creation of measurement standards and new methods of measurements is entrusted to the State Metrology Service, which is under the jurisdiction of Gosstandart. Metrology is of great importance for standardization and unification of technological processes and products.

2.1.1 Basic terms of metrology

A very important factor of correct understanding of the discipline and science of metrology are the terms and concepts used in it. It is necessary to say that their correct formulation and interpretation are of paramount importance, because the perception of each person is individual and many, even generally accepted terms, concepts and definitions he interprets in his own way, using his life experience and following his instincts, his life credo. And for metrology it is very important to interpret terms unambiguously for all, because this approach gives an opportunity to understand any life phenomenon optimally and completely. A number of special terms are used in metrology.

A *physical* quantity representing a common property with respect to the quality of a large number of physical objects, but individual for each in the sense of quantitative expression.

A *unit of physical quantity*, which implies a physical quantity to which a numerical value equal to one is assigned by convention.

Measurement - finding the values of a physical quantity experimentally with the help of special technical means. The basic equation of measurement is as follows:

$$Q = qU,$$

where Q is the value of the physical quantity;

q - numerical value of the quantity in the accepted units;

U is a unit of physical quantity.

The experiment produces a measured value of a quantity, i.e., a value of the quantity that approximates its true size (i.e., the true value of the quantity).

Measurement of physical quantities, which refers to the quantitative and qualitative evaluation of a physical object by means of measuring instruments.

A *measuring instrument* that is a technical *means* having standardized metrological characteristics. It includes a measuring instrument, a measure, a measuring system, a measuring transducer, a set of measuring systems.

A measuring instrument is a measuring instrument that produces an information signal in a form that is intelligible for direct perception by an observer.

A *measure* is also a means of measurement that reproduces a physical quantity of a given size.

A *measuring system*, perceived as a collection of measuring instruments that are interconnected by means of information transmission channels to perform one or more functions.

A measuring *transducer is* also a measuring instrument that produces an information measurement signal in a form convenient for storage, viewing and broadcasting via communication channels, but not available for direct perception.

A *measurement principle* is a physical phenomenon or a set of physical phenomena on which a measurement is based. For example, temperature measurement is based on the phenomenon of liquid expansion when heated (mercury in a thermometer). Principle of measurement as a set of physical phenomena on which the measurement is based.

Measurement method as a set of techniques and principles of using technical measuring instruments.

A unit standard is a measure or measuring instrument designed to reproduce a physical quantity on a national or international scale. There are standards of kilogram, ampere, second, etc. (more than a hundred primary and special standards).

Working measuring instrument - a measure or measuring instrument intended for making technical *measurements*.

A standard sample is a measure for reproducing units of quantities characterizing the properties or composition of substances and materials.

A reference material is a measuring instrument in the form of a substance (material), the composition or property of which is reliably established by attestation. They are used for calibration, attestation and verification of measuring instruments, control of correctness of measurement results.

There is a distinction between standard samples of composition and standard samples of properties, with the latter serving as measures. Thus, for verification of dilatometers (devices measuring temperature deformations of materials), standard samples of properties are used in the form of reference measures made of ultrapure copper (for temperatures from -100 to +100), crystalline quartz (t=20-5000C), corundum (temperatures below 900^0 C).

Measurement accuracy is a characteristic that expresses the degree to which the measurement results correspond to the present value of the measured quantity. Quantitatively, the accuracy of measurement is equal to the value of the relative error in minus the first degree taken modulo the first degree.

Correctness of measurement is a qualitative characteristic of measurement, which is determined by how close to zero is the value of constant or fixedly varying error (systematic error). This characteristic depends, as a rule, on the accuracy of measuring instruments.

Reliability of measurements is a characteristic that determines the degree of confidence in the obtained measurement results. According to this characteristic, measurements are divided into reliable and unreliable. Reliability of measurements depends on whether the probability of deviation of measurement results from the present value of the measured quantity is known. If the reliability of measurements is not determined, the results of such measurements, as a rule, are not used. The validity of measurements is limited from above by the measurement error.

Measurement error is the difference between the result of measurement of a quantity and the present (actual) value of this quantity. The error usually arises due to insufficient accuracy of measuring instruments and methods or due to the impossibility to ensure identical conditions under repeated observations.

Measurement error:

$$\delta = x - X,$$

where x is the measured value of the quantity;

X is the true value of the quantity.

The true value of the measured quantity always remains unknown due to the lack of ideal methods and means of measurement, so in practice, instead of the true value, the result of measurement obtained by more accurate methods and means is used and called the *actual value*. Thus, the value of δ is determined with some approximation to the true value.

The error of measurement can be expressed in units of the measured quantity (*absolute error*) or in fractions, percent of its value (*relative error*). The error of measures and instruments is determined by their verification.

Verification is a set of actions to assess the errors of measures and measuring instruments.

2.2 Types and methods of measurements

Classification of measuring instruments may be carried out according to the following criteria.

1. According to the characteristic of measurement accuracy, measurements are

divided into equally accurate and unequally accurate.

Equally accurate measurements of a physical quantity are a series of measurements of some quantity made with measuring instruments (MI) of the same accuracy under identical initial conditions.

Unequal measurements of a physical quantity are a series of measurements of some quantity made with measuring instruments of different accuracy and (or) under different initial conditions.

2.   According to the number of measurements, measurements are divided into single and multiple measurements.

A single measurement is a measurement of one quantity taken once. In practice, single measurements have a large error, therefore it is recommended to perform at least three measurements of this type and take the arithmetic mean as the result to reduce the error.

A repeated measurement is a measurement of one or more quantities made four or more times. A repeated measurement is a series of single measurements. The minimum number of measurements at which a measurement can be considered repeated is four. The result of a repeated measurement is the arithmetic mean of the results of all measurements made. Repeated measurements reduce the error.

3.   According to the type of change in the measurement value, measurements are divided into static and dynamic.

Static measurements are measurements of a constant, unchanging physical quantity. An example of such a constant physical quantity in time is the length of a land plot.

Dynamic measurements are measurements of a changing, non-constant physical quantity.

4.   According to the purpose of measurements are divided into technical and metrological.

Technical measurements are measurements performed by technical measuring instruments.

Metrological measurements are measurements made using standards.

5.   According to the way the measurement result is presented, they are divided into absolute and relative.

Absolute measurements are measurements that are made by direct, immediate measurement of a basic quantity and/or application of a physical constant.

Relative measurements are measurements in which the ratio of homogeneous quantities is calculated, with the numerator being the quantity to be compared and the denominator being the base of comparison (unit). The result of the measurement will depend on which quantity is taken as the base of comparison.

6.   According to the methods of obtaining the results of measurements are divided into direct, indirect, joint and aggregate.

Direct measurements are measurements made with the help of measures, i.e. the quantity to be measured is compared directly with its measure. An example of a direct measurement is the measurement of an angle (the measure is a protractor).

Indirect measurements are measurements in which the value of the measured quantity is calculated using the values obtained by direct measurements and some known relationship between these values and the measured quantity. For example, determination of the volume of bodies of correct geometric shape by the results of direct measurements of its linear dimensions and the corresponding mathematical calculation. The same applies to the determination of the density of materials, compressive strength.

Joint measurements are measurements during which at least two heterogeneous physical quantities are measured in order to establish the existing dependence between them. In this case, the values of the measured quantities are found from the data of repeated direct or indirect measurements of the quantities that are not similar.

Repeated measurements are carried out with different combinations of measures or under changing conditions, which makes it possible to make a system of equations, solving which, the desired value of the measured value is found. This method is used, for example, in determining the modulus of elasticity of concrete.

Cumulative are measurements of several quantities of the same name made simultaneously, in which the desired values are found by solving a system of equations obtained by direct measurements of various combinations of these quantities.

Varieties of direct measurements:

- direct assessment method;

- differential method;

- null method;

- coincidence method.

The *method of direct estimation* allows you to obtain the value of a value directly, without any additional actions and without calculations (exception - multiplication of the reading by the constant of the device or by the price of division). Such measurements are made on manometers, dynamometers, liquid thermometers, weighing on dial scales, measuring length with a ruler).

The *differential (difference) method* consists in measuring the difference between the measured quantity and a quantity whose value is known.

The *zero method* consists of comparing the measured quantity with a quantity whose value is known in advance. Both quantities are chosen to be equal in size, so that the difference between them will be zero. This method is used to determine mass on a lever scale when the mass of the weights is selected equal to the mass to be measured. This method is similar to the differential method, but in the zero method the difference is reduced to zero.

The *coincidence method* is to measure from coincident marks or signals [11].

2.3 Measurement errors

They inevitably occur during measurements. They consist of instrumental error and measurement method error, which may have systematic and random components. In addition, errors or gross errors may occur during measurement.

Systematic errors are errors that remain constant or change according to a well-defined law during successive measurements. Systematic errors can be studied and the measurement result can be clarified by introducing corrections to the readings of measuring devices.

Random errors are those errors that take different values when measuring the same quantity.

Theoretical errors are related to the error of the measurement method itself. For example, when measuring the volume of bodies, their shape is assumed to be geometrically correct, so the dimensions are measured in an insufficient number of places.

Subjective errors are the result of individual qualities of a person due to the peculiarities of his senses or acquired incorrect measurement skills (counting the volume in a burette by the lower meniscus).

Ways to exclude and account for systematic errors:

1) elimination of sources of errors before the start of measurements (error prevention);

2) elimination of errors in the measurement process (experimental elimination of errors);

3) making known corrections to the measurement result (elimination of errors by calculation);

4) estimation of the limits of systematic errors that cannot be excluded

5) **4 The concept of a physical quantity. Units of measurement**

Physical quantity is a concept of at least two sciences: physics and metrology. By definition, *a physical* quantity is a certain property of an object, process, common to a number of objects in terms of qualitative parameters, but differing, however, in quantitative terms (individual for each object). A classic example of illustration of this definition is the fact that, having their own mass and temperature, all bodies have individual numerical values of these parameters. Accordingly, the size of a physical quantity is considered to be its quantitative filling, content, and in turn, the value of a physical quantity is a numerical assessment of its dimensions.

All values of physical quantities are traditionally divided into true and actual values. True values are values that ideally reflect qualitatively and quantitatively the corresponding properties of the object, and actual values are values found experimentally and are so close to the truth that they can be accepted instead of it. However, the classification of physical quantities is not exhausted by this. There are a number of classifications created on various grounds. The main ones are the divisions:

1)   into active and passive physical quantities - by division in relation to measurement information signals. *Active physical* quantities are quantities that have a probability to be transformed into a measuring information signal without using auxiliary energy sources. And *passive* ones represent such quantities, for measurement of which it is necessary to use auxiliary energy sources, which create a signal of measuring information;

2)   additive (or extensive) and non-additive (or intensive) physical quantities - when divided on the basis of additivity. It is considered that *additive* quantities are measured in parts, besides, they can be accurately reproduced with the help of a multivalued measure based on summation of dimensions of separate measures. And *non-additive* quantities are not directly measurable, as they are transformed into a direct measurement of a quantity or measurement by indirect measurement.

In 1791, the French National Assembly adopted the first ever system of units of physical quantities. It was a metric system of measures. It included: units of lengths, areas, volumes, capacities and weights. And they were based on two now commonly known units: meter and kilogram. A number of researchers believe that, strictly speaking, this first system is not a system of units in the modern sense. It was only in 1832 that the German mathematician K. Gauss developed and published the latest methodology for building a system of units, which in this context is a certain set of basic and derived units.

The scientist based his methodology on three basic quantities independent of each other: mass, length, time. And as the basic units of measurement of these quantities, the mathematician took milligram, millimeter and second, since all other units of measurement can be easily calculated with the help of minimum units. K. Gauss considered his system of units to be an absolute system. With the development of civilization and scientific and technological progress, a number of other systems of units of physical quantities have emerged, the basis for which is the principle of the Gauss system. All these systems are built as metric, but they are distinguished by different basic units. Thus, at the present stage of development, the following basic systems of units of physical quantities are distinguished:

1)   GHS system (1881) or the GHS system of units of physical quantities, the basic units of which are the following: the centimeter (cm) - represented as a unit of length, the gram (g) - as a unit of mass, and the second (s) - as a unit of time;

2)   The ICGSS system (late 19th century), initially using the kilogram as a unit of weight and later as a unit of force, which led to the creation of a system of units of physical quantities, the basic units of which became three physical units: the meter as a unit of length, the kilogram - force as a unit of force, and the second as a unit of time;

3)   the ICCA system (1901), the foundations of which were created by the Italian scientist G. Giorgi, who proposed the meter, kilogram, second and ampere as the units of the ICCA system.

Today in the world science there is an incalculable number of all kinds of systems of units of physical quantities, as well as many so-called extra-system units. This, of course, leads to certain inconveniences in calculations, forcing to resort to recalculation when translating physical quantities from one system of units to another. A situation arose in which there was a serious need to unify the units of measurement. Such a project was created in 1954 by the Commission for the Development of a Unified International System of Units. It was called the "Draft International System of Units" and was eventually approved by the General Conference on Weights and Measures. Thus, a system based on seven basic units became known as the International System of Units, or SI for short.

Decisions of the General Conference on Weights and Measures have adopted such definitions of the basic units of measurement of physical quantities:

1)   meter is considered to be the length of the path that light travels in a vacuum in 1/299,792,458th of a second;

2)   kilogram is considered to be equated to the existing international prototype of the kilogram;

3)   second is equal to 919 2631 770 periods of radiation corresponding to the transition that occurs between the two so-called superfine levels of the ground state of the Cs133 atom;

4)   The ampere is considered to be a measure of that force of unchanging current which causes on each section of a conductor of 1 m length a force of interaction under the condition of passing through two rectilinear parallel conductors having such parameters as negligibly small circular cross-sectional area and infinite length, as well as location at a distance of 1 m from each other under vacuum conditions;

5)   kelvin is equal to 1/273.16th of the thermodynamic temperature, the so-called triple point of water;

6)   mol is equal to the amount of matter of a system that contains the same number of structural elements as the atoms in C 12 of mass 0.012 kg.

In 1960, the XI General Conference on Weights and Measures approved the International System of Units (SI).

The International System of Units is based on seven units covering the following fields of science: mechanics, electricity, heat, optics, molecular physics, thermodynamics and chemistry:

1)   The unit of length (mechanics) is the meter;

2)   The unit of mass (mechanics) is the kilogram;

3)   The unit of time (mechanics) is the second;

4)   The unit of electric current force (electricity) is the ampere;

5)   The unit of thermodynamic temperature (heat) is the kelvin;

6)   The unit of light intensity (optics) is the candela;

7)   The unit of quantity of matter (molecular physics, thermodynamics and chemistry) is the mole.

8)   The International System of Units has additional units:

1   ) The unit of measurement of a plane angle is the radian;

2   ) The unit of measurement of the solid angle is the steradian. Thus, by adopting the International System of Units, the units of measurement of physical quantities in all fields of science and technology were organized and brought to one form, since all other units are expressed through the seven basic and two additional SI units. For example, the quantity of electricity is expressed in seconds and amperes[12].

3   .5 Physical quantities and measurements

The object of measurement for metrology, as a rule, are physical quantities. *Physical quantities are* used to characterize various objects, phenomena and processes. They are divided into basic and derived from the basic quantities. Seven basic and two additional physical quantities are established in the International System of Units. These are length, mass, time, thermodynamic temperature, quantity of matter, light intensity and electric current, additional units are radian and steradian.

Physical quantities have qualitative and quantitative characteristics. *Qualitative difference of physical quantities* is reflected in their dimension. The designation of dimensionality is established by the international standard ISO, it is the symbol dim*.

An indicator of the degree of dimensionality can take different values and different signs, can be both integer and fractional, can take the value zero. If, when determining the dimensionality of a derived quantity, all degree indices are equal to zero, then the base of the degree, respectively, takes the value of one, thus, the quantity is dimensionless.

The *quantitative* characteristic of a measurement object is its size obtained as a result of measurement. The most elementary way to obtain information about the size of a certain value of a measurement object is to compare it with another object. The result of such a comparison will not be an exact quantitative characteristic, it will only allow to find out which of the objects is larger (smaller) in size. Not only two dimensions can be compared, but also a larger number of dimensions. If the sizes of the measurement objects are arranged in ascending or descending order, the scale of order will be obtained. The process of sorting and arranging the sizes in ascending or descending order on the order scale is called ranking. For ease of measurement, certain points on the order scale are fixed and are called reference points. The fixed points on the order scale can be assigned numbers, which are often referred to as scores.

4 .6 Standards and exemplary measuring instruments

All issues related to storage, application and creation of standards, as well as control over their condition, are solved according to the unified rules established by GOST. Standards are classified according to the principle of subordination. According to this parameter, standards are primary and secondary.

Primary measurement standards reproduce and/or store units and transfer their dimensions with the highest accuracy available in the given field of measurement. A primary standard shall serve the purpose of ensuring the reproduction, storage of units and transmission of dimensions with the highest accuracy that can be obtained in a given field of measurement. In turn, primary standards may be special primary standards, which are intended for reproduction of the unit in conditions when direct transfer of the unit size with the required accuracy is practically impossible. They are approved in the form of state standards. Since there is a special significance of state standards, any state standard is approved by GOST. Another task of this approval is to give these standards the force of law. The State Committee on Standards has the responsibility to create, approve, store and use state standards.

Secondary measurement *standards* are a transfer link between the primary standard and working measuring instruments. Their accuracy is lower than that of the primary standard. The secondary standard reproduces the unit under special conditions, replacing the primary standard under these conditions. It shall be created and approved for the purpose of ensuring minimum deterioration of the state standard. Secondary measurement standards can be divided on the basis of purpose. Thus, there are:

1)   copy standards intended for transferring unit sizes to working standards;

2)   comparison standards intended to verify the integrity of the State Standard and for the purpose of its replacement in case of its deterioration or loss;

3)   witness standards intended for the comparison of standards which, for a number of different reasons, are not directly comparable with each other;

4)   working standards, which reproduce the unit from secondary standards and serve to transfer the size to a standard of a lower grade. Secondary standards are created, approved, maintained and used by ministries and agencies.

There is also the notion of "*unit standard*", which means a single instrument or a set of instruments aimed at reproducing and storing a unit for the subsequent transmission of its size to inferior instruments, made to a specific specification and officially approved in the prescribed manner as a standard. There are two ways of reproduction of units on the basis of technical and economic requirements:

1)   centralized method - with the help of a single state standard for the whole country or a group of countries. All basic units and most of the derivatives are reproduced centrally;

2)   decentralized method of reproduction - applicable to derived units, information about the size of which is not transferred by direct comparison with the standard [13].

Size translation can occur by different methods of verification. As a rule, size transmission is performed by known measurement methods. On the one hand, there is a certain disadvantage of the step size transmission, which implies that sometimes there is a loss of accuracy. On the other hand, there are positive aspects, which imply that this multistage method helps to protect the standards and to transfer the size of the unit to all working measuring instruments. There is also a notion of "exemplary measuring instruments", which are used for regular translation of unit sizes in the process of verification of measuring instruments and are used only in metrological service units. The category of an exemplary measuring instrument is determined in the course of metrological certification by one of the bodies of the State Committee on Standards. If necessary, particularly accurate

working measuring instruments may be certified for a specified period as reference measuring instruments in the above-mentioned order. Conversely, exemplary measuring instruments that have not passed the next certification for various reasons shall be used as operational measuring instruments.

2.7 Measuring instruments and their characteristics

In scientific literature, technical measuring instruments are divided into three large groups. These are: measures, calibers and universal measuring instruments, which include measuring instruments, control and measuring instruments (CMP), and systems.

1.   A *measure* is a measuring instrument that is designed to reproduce a physical quantity of a proper size. Measures include plane-parallel length measures (tiles) and angle measures.

2.   *Calibers* are devices whose purpose is to be used for controlling and finding the dimensions, interposition of surfaces and shape of workpieces within the desired boundaries. As a rule, they are divided into: smooth limit gauges (staples and plugs), and threaded gauges, which include threaded rings or staples, threaded plugs, etc.

3.   A *measuring instrument* represented as a device that produces a measurement information signal in a form understandable to observers.

4.   *Measuring system* understood as a certain set of measuring instruments and some auxiliary devices, which are connected with each other by communication channels. It is intended for the production of measurement information signals in some form suitable for automatic processing, as well as for transmission and application in automatic control systems.

5.   *Universal measuring instruments*, the purpose of which is to be used to determine the actual dimensions. Any universal measuring instrument is characterized by its purpose, principle of operation, i.e. the physical principle underlying its construction, design features and metrological characteristics [14].

2.8   Classification of measuring instruments

A measuring instrument (MI) is a technical means or a set of means used for making measurements and having standardized metrological characteristics. With the help of measuring instruments a physical quantity can be not only detected but also measured.

Measuring instruments are classified according to the following criteria:

1)   on the ways of constructive realization;

2)   for metrological purposes.

Measuring instruments are divided according to the way of constructive realization:

1)   on measures of magnitude;

2)   measuring transducers;

3)   measuring devices;

4)   measuring installations;

5)   measurement systems.

Measures of magnitude are means of measurement of a certain fixed size, repeatedly used for measurement. Distinguished:

1)   unambiguous measures;

2)   multivalued measures;

3)   measure sets.

2.9 Metrological characteristics of measuring instruments

Metrological properties of measuring instruments are properties that have a direct influence on the results of measurements performed by these instruments and on the uncertainty of these measurements.

Quantitative and metrological properties are characterized by indicators of metrological properties, which are their metrological characteristics.

Metrological characteristics approved by ND are standardized metrological characteristics. Metrological properties of measuring instruments are subdivided:

1)   on the properties that define the scope of application of measuring instruments:

2)   properties that determine the precision and accuracy of the measurement results obtained.

The properties that establish the scope of application of measuring instruments are defined by the following metrological characteristics:

1)   measurement range;

2)   sensitivity threshold.

The range of *measurement* is the range of values of a quantity in which the error limits are normalized. The lower and upper (right and left) limit of measurement is called the lower and upper limit of measurement.

The *sensitivity threshold* is the minimum value of the measured quantity that can cause a noticeable distortion of the received signal.

The properties that determine the precision and correctness of the obtained measurement results are determined by the following metrological characteristics:

1)   the correctness of the results;

2)   precision of results.

3)   10 Basic measurement characteristics

The following main characteristics of measurements are distinguished:

1)   the method by which the measurements are made;

2)   measurement principle;

3)   measurement error;

4)   measurement accuracy;

5) correctness of measurements;

6) measurement reliability.

Measurement method is a method or a set of methods by means of which a given quantity is measured, i.e. a comparison of the measured quantity with its measure according to the accepted principle of *measurement.*

There are several criteria for classifying measurement methods.

1. According to the methods of obtaining the desired value of the measured quantity, they are distinguished:

1) direct method (carried out by direct, direct measurements);

2) indirect method.

2. According to the methods of measurement are distinguished:

1) contact measurement method;

2) non-contact method of measurement. The contact method of measurement is based on direct contact of any part of the measuring instrument with the object to be measured.

With the non-contact measurement method, the measuring instrument does not directly contact the object to be measured.

3. According to the methods of comparing a value with its measure, they are distinguished:

1) direct assessment method;

2) a method of comparison with her unit.

The direct estimation method is based on the use of a measuring instrument that shows the value of the measured quantity.

The measure-to-measure method is based on comparing a measurement object with its measure.

2.11 Metrological support, its basics

Metrological support means a set of actions to achieve unity and required accuracy of measurements. Metrological support is based on three bases: scientific, technical and organizational.

The *scientific basis of* metrological support is metrology. The system of state and working standards of units of physical quantities, working measuring instruments, standard samples of composition and properties of substances and materials, standard reference data on physical constants and properties of substances and materials, as well as the system of obligatory state and departmental verification and certification of measuring instruments shall constitute the *technical basis of* metrological support.

The *organizational basis* is the metrological service of the country consisting of state and departmental metrological services.

Unity of measurements in the country is controlled by Gosstandart through metrological institutions and Gosnadzor bodies. Its system includes verification laboratories performing the service of Gosnadzor, verifications and state tests. They ensure reliability and unity of measurements by means of state verification of the most accurate and responsible measuring and testing instruments, carry out state tests of measures and measuring instruments. Unity of measurements in branches of national economy is controlled by special subdivisions of branch metrological service, which use charters of Gosnadzor bodies of Gosstandart.

Metrological support, or abbreviated as *ME*, is the establishment and use of scientific and organizational bases, as well as a number of technical means, norms and rules necessary for observance of the principle of unity and required accuracy of measurements. Nowadays the development of MI is moving in the direction of transition from the existing narrow task of ensuring the unity and required accuracy of measurements to the new task of ensuring the quality of measurements. The meaning of the notion "metrological assurance" is deciphered in relation to measurements (testing, control) in general. However, this term is also applicable in the form of the concept "metrological assurance of technological process (production, organization)", which implies MO of measurements (testing or control) in this process, production, organization. All stages of the life cycle (LC) of a product (product) or service can be considered as the object of MI, where the life cycle is perceived as a certain set of successive interrelated processes of creation and change of the product state from the formulation of initial requirements to the end of operation or consumption. Often at the stage of product development to achieve high quality of the product is the choice of controlled parameters, accuracy standards, tolerances, means of measurement, control and testing. And in the process of development of MI it is desirable to use a systematic approach, in which the specified provision is considered as a certain set of interrelated processes, united by one goal. This goal is to achieve the required quality of measurements. As a rule, scientific literature distinguishes a number of such processes:

3) Establishing the nomenclature of the parameters to be measured and the most appropriate accuracy standards for product quality control and process control;

4) feasibility study and selection of measuring, testing and control equipment and establishment of their rational nomenclature;

5) standardization, unification and aggregation of used control and measuring equipment;

6) development, implementation and certification of modern methods of measurement, testing and control (MBI);

7) verification, metrological attestation and calibrations of instrumentation or control and measuring and testing equipment used at the enterprise;

8) control over the production, condition, application and repair of KIOs, as well as over the exact adherence to the metrology rules and norms at the enterprise;

9) Participation in the process of creating and implementing enterprise standards;

10) implementation of international, state and industry standards, as well as other regulatory documents of Gosstandart;

11) carrying out metrological examination of design, technological and normative documentation projects;

12) analyzing the state of measurement, developing on its basis and implementing various activities to improve the IO;

13) training of employees of relevant services and departments of the enterprise to perform control and measurement operations.

1.12 State metrological supervision. Metrological service

The metrological service of the Republic of Kazakhstan consists of:

- from the state metrological service;

- state time and frequency services; standard samples of composition and properties of substances and materials; standard reference data on physical constants and properties of substances and materials;

- metrological services of management bodies, individuals and legal entities.

The metrological service is an important part of public administration and solves the following tasks:

- implementation of a set of measures on metrological support of business entities' activities;

- ensuring unity and accuracy of measurements;

- ensuring safety, quality and increasing the competitiveness of domestic products in the global market;

- ensuring reliable accounting of all types of material and energy resources;

- protection of interests of citizens and economy of the Republic of Kazakhstan from consequences of unreliable measurement results.

The State Metrology Service is headed by Gosstandart RK. It consists of:

- state scientific and metrological center, which ensures creation, improvement, storage and application of state measurement standards of units of magnitudes, creation of systems for transfer of dimensions of units of magnitudes, development of normative documents to ensure the uniformity of measurements;

- territorial bodies of Gosstandart, which perform on the respective territory tasks and functions on assuring the uniformity of measurements within the limits defined by the provisions on these subdivisions.

Gosstandart RK carries out the following functions in the field of metrology:

- forms and implements the state policy on ensuring the uniformity of measurements;

- carries out coordination of activity of metrological service of RK;

- establishes the units of quantities to be used;

- organizes fundamental research in the field of metrology;

- establishes rules for creation, approval, storage and use of state measurement standards of units of magnitudes, improves the standard base of units of magnitudes of the Republic of Kazakhstan;

- defines general metrological requirements for measuring instruments, methods and results;

- represents the Republic of Kazakhstan in international and regional organizations on metrology;

- organizes professional training and retraining of personnel in the field of assurance of measurement uniformity.

In accordance with the tasks the main range of responsibilities of metrological services of legal entities includes:

- systematic analysis of the state of measurements, control and testing at all stages of development, production and operation of certain types of products;

- participation in the development of measuring instruments and methods and their implementation;

- participation in creation of measurement standards, other verification means necessary for metrological service of created and manufactured measuring instruments;

- preparation of materials and participation in coordination of applications for import of measuring instruments;

- organization and carrying out of works on verification of measuring instruments, ensuring timely submission of measuring instruments for verification;

- keeping records of measuring instruments, coordination and observance of calibration schedules of measuring instruments, timely writing off obsolete measuring instruments;

- realization of rental of measuring instruments;

- making particularly precise measurements;

- development and approval of local verification schemes;

- implementation of metrological supervision over the condition and application of measuring instruments, certified measurement methods, measurement standards.

Measuring instruments to be manufactured or imported from abroad shall be subjected to state tests followed by type approval or metrological certification [15].

Carrying out of state tests, type approval and entering into the state register of measuring instruments is carried out by Gosstandart RK.

The manufacturer is obliged to affix the state register mark on the type-approved instrumentation or in its operational documentation.

In accordance with concluded international treaties, the results of state tests, verification, metrological certification of measuring instruments conducted in foreign countries are recognized.

For violation of metrological rules and norms Gosstandart has the right to prohibit the use of measuring instruments that have not passed state tests and type approval, verification or metrological certification, do not conform to the approved type, to draw up protocols on administrative responsibility.

PR RK 50.2.21-95 "Procedure of state tests and type approval of measuring instruments" establishes general requirements for organization and procedure of state tests and type approval of measuring instruments. The procedure applies to measuring instruments produced in the territory of the Republic and imported from abroad, and is developed in development of the Law of RK "On Ensuring Uniformity of Measurements".

Test procedures include:

- testing of MIs for the purposes of type approval;

- taking a decision on type approval, its state registration and issuance of a type approval certificate;

- testing of MIs for conformity to the approved type when controlling conformity of MIs to the approved type;

- recognition of type approval or results of tests of SI type carried out by competent organizations of foreign countries;

- information service for consumers of measuring equipment.

Type approval is a type of state metrological supervision and is carried out in order to ensure uniformity of measurements in the Republic.

Decision on type approval is made by Gosstandart RK based on the results of testing of MI for the purposes of type approval. Applications for conducting tests of measuring instruments for type approval shall be sent to Gosstandart RK in the prescribed form.

During testing of MI for the purposes of type approval, compliance of the technical documentation and technical characteristics of MI with the requirements of the technical specification, technical conditions and normative and operational documents applicable to them, including methods of verification of MI shall be checked.

In case of positive test results, Gosstandart RK makes a decision on type approval of SI, which is certified by a type approval certificate.

Measuring instruments for which type approval certificates have been issued are subject to registration in the State Register of Measuring Instruments.

The applicant shall affix the type approval mark to the measuring instruments whose type has been approved and to the operational documentation accompanying each instrument. If it is not expedient to affix the mark on the measuring instrument, it is allowed to affix it only on the operational documents.

According to international agreements concluded by the Government, results of tests and type approval of other countries can be recognized in the Republic, which is the basis for entering the type of imported measuring instruments into the State Register and their application in the Republic.

1.13 Standardization of measurement methods and instruments in the field of building materials

Standardized methods and means of measurement designed to determine: material composition (chemical, mineralogical, phase); material structure (solid matter, pore space); quality indicators established by the standard of technical specifications for the material.

Quality indicators can be: physical quantities with appropriate dimensions (density, thermal conductivity, etc.); technical characteristics measured in conventional units (water resistance, frost resistance, etc.). To determine quality indicators, physical methods based on the laws of physics are used, as well as comparative methods of measuring technical characteristics in conventional units (freezing and thawing cycles, etc.).

Standardization of measuring instruments is possible only after their state tests, which include examination of technical documentation for newly developed measuring instruments and their experimental study, conducted by the bodies of the State Metrological Service or on their behalf [16].

Standardize the following basic provisions of the methodology for measuring the mechanical properties of building materials:

- shape and size of the specimen, allowable deviations in shape and dimensions;

- number of samples to be tested simultaneously to determine the property values;

- method of taking an average sample of material and making samples from it;

-       equipment used for testing and determining properties;

-       the loading pattern of the specimens;

-       preparation of the specimen for testing (flatness, presence of defects);

-       test conditions (temperature and humidity of the specimen and environment);

-       processing of measurement results and determination of material property index (strength grade, etc.) taking into account the scale factor.

3 Basic provisions of product quality

3.1 The concept of quality

In accordance with international standards, *quality* is understood as a set of properties and characteristics of a product or service that give it the ability to satisfy a specified or intended need.

Products are considered as a materialized result of labor activity, possessing useful properties and intended to meet the needs of both public and personal nature. The products of construction are buildings and structures, and the products of the construction industry are materials and products.

Materials are non-piece products whose quantities are expressed using continuous quantities. For example: mass of cement produced, area of window glass, volume of concrete mix. Materials also include small piece products produced in large quantities: bricks (thousand pieces), ceramic facing tiles (thousand m2), etc.

Products are industrial products, the quantity of which is calculated by a certain number of copies (pieces). They represent a pre-manufactured element or part of a construction. In some cases, construction products - reinforced concrete, metal, wood - are called *structures*.

Each type of product exhibits quite specific properties of interest to the consumer. For construction industry products, these are strength and density of material, degree of dimensional accuracy of products, thermal conductivity, frost resistance, resistance to water, aggressive liquids and gases.

The technical and economic concept of product quality does not include all properties, but only those that are associated with the satisfaction of certain personal or social needs. The following concepts are used to characterize the properties: product quality indicator, attribute and product parameter.

Product quality indicator is a quantitative characteristic of one or more properties that constitute product quality, considered in relation to certain conditions of creation and operation (consumption). The nomenclature of quality indicators depends on the purpose of the product.

A product feature is a qualitative or quantitative characteristic of any properties or states of the product, a *parameter* is a quantitative characteristic of properties.

Quality attributes are the color and shape of the product, the presence of protective and decorative coating on its surface, etc. They are used in alternative control and management of product quality.

A *product parameter* quantitatively characterizes any of its properties, including those that are part of quality.

In qualitative assessment, any product property is necessary and sufficiently defined by three numerical parameters: size (absolute indicator), score (relative indicator), and weighting.

The dimension of a property is determined by measuring the physical, mechanical and other characteristics of the material and is expressed in appropriate units.

For example, the strength dimension of concrete is its tensile strength in megapascals (MPa).

Evaluation characterizes the degree of satisfaction of social need in a given property. In the process of evaluation, the value of any product indicator is compared with the base or reference indicator.

The weighting determines the importance of a given property among the other properties that make up the quality. The sum of the weights of all properties is a constant value. Increasing the weight of one of the properties can occur at the expense of decreasing the weight of the others.

Quality is a complex technical and economic category, it is variable and unstable, subject to the influence of many technical, organizational and economic factors. Thus, some quality indicators of construction products change under the influence of climatic factors. Moistening of exterior wall elements leads to an increase in their thermal conductivity, reducing indoor comfort [17].

Quality assessment methods should take into account the change of product parameters in the process of its creation and operation. This can be achieved by using a set of quantitative indicators: single, complex and integral.

A single quality indicator characterizes one of the product properties, e.g. mobility of a concrete mix. A single indicator can refer both to a single product unit and to a group of homogeneous products, characterizing one property.

A complex quality indicator characterizes jointly several simple properties or one complex property of the product consisting of several simple ones.

The division of quality indicators into unit, complex and integral ones makes it possible to solve many practical problems and, in particular, to quantitatively measure the quality of products.

Qualimetry is the measurement of product quality. It is a branch of metrology that studies the measurement of quality.

3.1.1 Assignment indicators

Purpose indicators characterize the useful effect from the use of products for their intended purpose and determine the scope of its application. These indicators include: strength (compressive and tensile strength, rigidity, crack resistance, impact strength, seismic resistance), thermophysical indicators and resistance to external influences (frost resistance, moisture resistance, resistance to solar radiation, heat resistance, fire resistance, thermal conductivity, water resistance, sound insulation, light transmission, etc.).

When assessing the quality level of products, destination indicators are often used in conjunction with indicators of other types, such as reliability and durability, and sometimes with ergonometric and aesthetic indicators.

Reliability and durability indicators characterize the properties of reliability and durability of materials, products or construction objects.

Reliability indicators characterize the degree to which the product performs its functions provided that the rules of operation are followed.

The reliability property is laid down at the stage of product development, ensured at the stage of its production and maintained at the stage of operation.

Reliability is a complex property of a product, which consists of private properties: *durability, failure-free operation, maintainability and serviceability.*

Building products are categorized as repairable and non-repairable. *Repairable products* can be repaired or replaced in case of failure. *Non-recoverable* (wall panel connections, embedded parts, concealed wiring) cannot be repaired or replaced.

The most important technical states of construction products are: serviceability, malfunction and operability. *Serviceable* is the state of the object, in which it fully meets all the requirements of normative and technical documentation NTD both in respect of major and minor parameters. In the *faulty* state, the object does not meet at least one of the requirements of NTD in respect of major or minor parameters. *Operable* is considered such a state of the object, in which the values of all the main parameters characterizing the ability to perform a given function, meet the requirements of NTD. Thus, serviceability necessarily includes operability, the reverse position is optional, i.e. a faulty object can be operable. For example, a serviceable object may not meet the aesthetic indicators.

It is necessary to distinguish the *limit state of the object,* in which its further use for its intended purpose is inadmissible due to physical deterioration or inexpedient due to moral deterioration. Criteria of the limit state are established in NTD. Thus, the criterion of frost resistance of concrete is the ultimate loss of strength (10-15%) or mass (5%) in cyclic freezing and thawing. In floor cladding materials, the limit state criterion is the relative mass loss during abrasion.

The transition of products (objects) from a serviceable state to a defective state occurs as a result of defects.

A defect is each individual non-compliance of products with the established requirements. Thus, if any unit of product has a defect, one or more parameters or indicators of its quality do not meet the requirements of NTD. Defects can be the output of the size of the product outside the tolerance limits, unacceptably high content of impurities (harmful) in the product, etc. Defects are divided into explicit and hidden. An explicit defect is a defect, for the detection of which in the NTD provides appropriate rules, methods and means of detection and control. Many defects are detected visually (cracks, splits, etc.); a number of defects are detected with the involvement of appropriate means of control - tools, instruments, fixtures. For example, deviation of actual brick dimensions from the specified dimensions is detected with the help of a ruler; surface curvature - with the help of an angle, ruler [18].

A latent defect cannot be detected by the methods, tools, etc. stipulated in the technical specifications. For example, hidden defects in ceramic bricks are foreign inclusions (stones, metal objects).

According to the degree of influence on the efficiency and safety of product use, critical, significant and insignificant defects are distinguished. *Critical* - a defect in which it is unacceptable to use the product for its intended purpose. *Significant* - a defect that significantly affects the performance of products and their durability, but is not critical. *Minor* - a defect characterizing such a deviation of a feature or parameter of the product, which does not significantly affect the use of the product for its intended purpose and its durability. The division of defects into critical, significant and insignificant defects is made for the purpose of selecting the type of product control - continuous or selective.

The same defect can be classified as avoidable or non-recoverable depending on at which stage of the process it is detected (early or final).

An unrecoverable defect is a defect, the elimination of which is technically impossible or economically unreasonable.

In the presence of defects, the object can be either in a state of damage or failure. *Damage* is an event consisting in the disturbance of the serviceable state of the object (its operability is preserved).

Failure characterizes an event consisting in the violation of the operable state of an object (it is assumed that the product was operable before the failure). The loss of operability is caused by such a malfunction, in which at least one of the main parameters exceeds the established tolerances.

Failures can be sudden and gradual. *Sudden failure - sudden* change of parameters (destruction); *gradual* (wear) - slow change of parameters (wear, deformation of material under the influence of environment).

The *reliability of* products, elements or systems consists of such essential properties as failure-free operation, durability, maintainability and serviceability.

Failure-free operation is the property of an object to continuously maintain a serviceable state for a certain time or a certain operating time (the latter for process equipment). MTBF is the duration or volume of work of an object from the beginning of its operation until the first failure occurs. It is possible to estimate the operating time by the amount of work performed.

The main indicators of manufacturability of industrial products include the coefficient of assemblability (blockiness) of the product and the coefficient of use of rational materials, as well as specific indicators of labor intensity of production, material and energy intensity of products.

The *coefficient of assemblability* (blockness) of the product characterizes the ease of assembly of the product and represents the share of structural elements included in the specified blocks in the total number of elements of the entire product.

The *coefficient of use of rational materials* is determined when it is expedient to use certain effective materials (polymeric, heat-insulating, composite) in the structure according to technical and economic indicators.

The manufacturability of products is characterized by the indicators of labor and material intensity. *Labor intensity of* production is determined by the amount of time spent on the manufacture of a unit of product, and is expressed for industrial products in *standard hours*.

Specific labor intensity is the ratio of the total labor intensity of production *T* to the main parameter of the product *B*. *Specific material intensity* - the ratio of the mass or finished product *M* to its main parameter *B*. Ergonomic quality indicators are used in determining whether the product meets the requirements of ergonomics (studies the interaction in the system man - environment - product).

Ergonomic indicators are divided into hygienic, anthropometric, physiological and psychological. The nomenclature of ergonomic quality indicators applies to industrial products, which include: workplace and interior equipment, monitoring and control panels, instruments and signaling devices, controls for technological equipment, industrial furniture. The level of ergonomic indicators is determined by ergonomics experts on a special rating scale in points.

Hygienic indicators characterize the compliance of the product with sanitary and household norms and recommendations. These indicators include illumination, temperature, humidity, pressure, electric and magnetic fields, dust, radiation, toxicity, noise, vibration, overload (acceleration).

Anthropometric indicators characterize products that are in direct contact with a person: elements of controls, clothing, footwear. Taking into account these indicators design, for example, control panels at automated stations of acceptance control of reinforced concrete products.

3.2 Standardization and unification indicators

They include indicators characterizing the degree of saturation of products with standardized and unified products.

The higher the quality of a product, the less it contains both original and standardized and unified parts.

Standardized are parts of the product manufactured according to state, national and industry standards; *unified* - manufactured according to enterprise standards, as well as received by the enterprise in finished form as component parts (from those in mass production). *Original* are the components designed specifically for this product.

The most important indicators of standardization and unification are the coefficient of applicability and the coefficient of repeatability.

The *applicability coefficient* characterizes the degree of saturation of the product with standardized and unified components.

A distinction is made between the applicability factor by size and the applicability factor by component part.

The degree of applicability of standardized components can also be expressed by means of a cost coefficient equal to the ratio of the cost of standardized components to the cost of the product as a whole. This coefficient can also be attributed to the group of economic indicators.

3.3 Methods of evaluation of product quality indicators

Quality assessment can be carried out by measuring, registration, calculation, organoleptic, expert and sociological methods. *Measuring* - determining the values of the product quality indicator with the help of technical means of measurement. This method is used to measure and control the majority of quality indicators of materials, products and structures: geometric dimensions, weight of products, strength, water absorption, etc. The basis of the measuring method is metrology.

Registration is based on the observation and counting of the number of certain events, items or costs. It is used to record product failures during testing, counting the number of defective products in a batch, etc.

When using the *design* method, calculations are made on the basis of established theoretical or empirical dependencies of product quality indicators on its parameters. This method is applied, as a rule, when designing products, when they can not yet be the object of experimental study. For example, determining the mass of a product by the values of its density and volume.

Organoleptic - determination of product quality based on the analysis of perception of human senses. This method is used to measure such properties of products that are not yet measurable by means of measuring instruments (assessment of the uniformity of color of facade ceramic products, the quality of the interior of the premises). The assessment is carried out by experts on the basis of their experience. Therefore, the reliability of the assessment depends on the qualifications of the expert.

The *expert* method of determining quality is based on the decision made by an expert. It is often used in forecasting the level of product quality. Expert methods are divided into individual and collective methods. Individual assessment is used when there is a very competent specialist in a given field of activity. Therefore, the method of collective expert evaluations based on the decision of a group of experts is more often used.

Sociological method is based on collection and analysis of opinions of actual or possible consumers of products. Opinions are collected by oral survey or by distributing questionnaires, holding conferences, exhibitions.

3.4 Quality control of materials and products

Improvement of product quality is based on a systematic approach that provides for systematic quality control and targeted impact on factors that can affect product quality. With this in mind, control, which should be carried out at the stages of product development, manufacturing and consumption, is of utmost importance.

Development of norms, methods and rules of quality control of products is one of the important areas of standardization. This includes: rules and norms of materials control, rules and methods of operational process control, rules of acceptance of finished products, methods of measuring product parameters, methods of testing products for resistance to external influences and reliability, requirements for testing and test equipment.

3.4.1 Types and methods of control

Control systems are classified by the level of conduct, objects of control, stages of the production process, the nature of the impact on product quality and other features.

Depending on the level of control, a distinction is made between: state, departmental and industrial control.

State control (supervision) over the quality of products is a set of technical and organizational measures to control the activities of enterprises, carried out by special state bodies: Gosstroy, Gosgrazhdanstroy, architectural and construction control bodies. The objects of control are not only the quality of products, but also the organization of the technological process, the state of equipment and NTD. As a result of such control reveals the degree of conformity of products to a given technical level for its certification.

Departmental control is carried out by the standardization services of ministries, institutes and trusts. It allows you to compare the activities of enterprises of this department and outline measures to improve the quality of products.

Production (technical) control is carried out by control bodies (technical control departments - GTC) of the enterprise. Depending on the controlled object, technical control is of the following types: labor control, product quality control and process control.

Labor control - verification of compliance of the labor results of employees engaged in production with the established requirements.

Product quality control - verification of compliance of product quality indicators with the established requirements.

Control of technological process - compliance of technological modes and other indicators of technological process with the requirements of STD.

Depending on the controlled stage of the production process, technical control is subdivided into incoming, operational and acceptance control.

Input control - determining the quality of raw materials, materials and component parts coming from supplier companies.

Operational control is carried out after the completion of certain technological operations.

Acceptance control is carried out to determine whether the quality of finished products conforms to the established requirements.

According to the completeness of coverage, the control can be continuous, selective, continuous, periodic and volatile. In the case of *continuous control,* the quality of all products without exception is determined. *Sampling control* consists in quality assessment based on the results of inspection of one or more samples or samples from a batch of products. A *batch* is a quantity of homogeneous products manufactured for a certain period of time from materials and semi-finished products of the same type and quality. The volume of batches is regulated by standards; the volume and number of samples - by control rules. As a rule, statistical methods are used for sample control.

Sampling inspection is appropriate to use when quality control is related to the product [19].

Continuous control is carried out to check the stability of the technological process. For this purpose, as a rule, automated means are used.

Flight control is carried out in special cases specified in the standards; its timeframe is not regulated.

According to the nature of impact on the production process, active and passive control are distinguished. *Active* control is carried out by measuring devices embedded in technological equipment and directly controlling changes in product properties; its results are used to control production processes.

Most of the existing control methods are carried out in a *passive* form, i.e. non-compliance of product quality indicators with the established requirements is detected when the defect can no longer be eliminated.

Traditional and *non-destructive testing methods* are distinguished by their impact on the object. In traditional inspection, quality information is obtained by laboratory methods, which mainly include destructive testing.

The main type of control is technical control. The objects of technical control are: material, semi-finished product, billet, part, assembly unit, complex, set, technological process.

The consumer performs incoming inspection in order to verify compliance of the quality of products coming to the enterprise with the requirements established in state standards, technical specifications, contracts. In the absence of data on the actual level of defectiveness of products, as well as with increased quality requirements, incoming inspection should be continuous, in other cases, selective. The results of incoming inspection are formalized in the form of cards.

In the course of operational control check the modes of preparation, placement and compaction of concrete mixture, the size and quality of the assembly of molds, the location of reinforcement and embedded parts, modes of heat and humidity treatment of concrete, the quality of finishing products.

Acceptance control consists in checking the compliance of finished products with the requirements of standards or specifications. The laboratory and QA check physical and mechanical properties of materials and products, assess the appearance and geometric parameters of products. The *release strength of* concrete in reinforced concrete products is assigned taking into account the conditions of transportation, installation and loading terms of products, as well as taking into account the technology of their manufacture.

3.4.2 Statistical control of product quality

Statistical current control consists of adjusting process parameters during production by means of random product control to ensure the required quality and prevent rejects. Thus, this control is as close as possible to the production process in terms of time. In the course of control, not all products are checked, but some part, a sample, the volume of which should be sufficiently representative (set by the standard depending on the volume of output).

In the process of control, it is checked whether the products by the controlled parameter have not exceeded the standardized tolerance, as well as determine the size of the actual deviation. If the controlled parameter is within the tolerance limits, the technological operation proceeds normally. If the parameter begins to approach the upper or lower limit of tolerance, there is a threat of defective products and even scrap. In this case, it is necessary to timely identify the causes of violation of technology and prevent marriage. Thus, statistical control is a kind of prophylactic means to prevent deterioration of product quality.

4 Basics of certification

4.1 Basic definitions of certification

The term "certification" in Latin means "done right", i.e. conformity confirmed. In essence, any conformity assessment is certification; all our activities are reduced to its three interrelated types: ordering and definition (standardization), control and measurement (metrology) and confirmation of results (certification).

Certification is a procedure by which a third party provides written assurance that a product, process or service meets specified requirements.

Compliance - the *conformity* of a product, process, or service with established requirements.

Third party - a person or body recognized as independent of the parties involved in the matter in question, (first party - suppliers, second party - buyers).

A *certification system* is a system that has its own rules of procedure and governance for carrying out conformity certification.

Certification Body - a body that conducts certification of conformity.

Certificate of Conformity - a document indicating that the necessary assurance is provided that a process or service complies with a specific standard or other (regulatory document) ND.

Conformity mark - a duly protected mark applied or issued in accordance with the rules of the certification system, indicating that the necessary confidence is provided that a given product or service conforms to a particular ND.

License (certificate) *in the field of certification* - a document by which a certification body grants a person or body the right to use certificates or marks of conformity for its products, processes or services in accordance with the rules of the relevant certification system.

Conformity assessment is the systematic verification of the extent to which a product, process or service conforms to specified requirements.

Accreditation - a procedure by which an authoritative body officially recognizes the eligibility of a person or body to perform specific work (produce products, provide services).

Certification of production - official confirmation by the certification body or other specially authorized body of the presence of the necessary conditions of production of products, ensuring the stability of requirements to it, specified in the ND.

Quality system - a set of organizational structure, responsibilities, procedures, processes and resources that provide overall quality management.

Product quality assessment and certification procedures are carried out by an independent, competent organization, such as a testing laboratory. To confirm its competence and objectivity, this organization must periodically undergo an accreditation procedure, i.e. official recognition of its ability to carry out the relevant type of control or testing.

Certification is standards-based and is based on testing against certification norms.

4.2 Aims and objectives of certification

The main objectives of certification are:

-	assisting consumers in competent choice of products (services);

-	protection of the consumer against unfairness of the manufacturer (seller, performer);

-	control of product (service, work) safety for the environment, life, health and property;

-	confirmation of product (service, work) quality indicators declared by the manufacturer (performer);

-	creation of conditions for the activity of organizations and entrepreneurs in the single commodity market of the RK, as well as for participation in international economic, scientific and technical cooperation and international trade.

-	confirmation of product conformity.

The very emergence of the concept of "confirmation of conformity" and filling it with modern meaning is associated with the recent sharp aggravation of the problem of quality of goods and services; globalization of international trade; a wide variety of products of the same functional purpose, but of different quality; fierce competition of commodity producers; the need to guarantee the safety of products for consumers.

Key Objectives:

-    ensuring the quality of work on certification of products and quality management systems of applicants in accordance with the rules and procedures established in the GOST Certification System;

-    fulfillment of the established requirements for Certification Bodies of products and quality systems;

-    observance of objectivity and impartiality in decision-making;

-    ensuring accessibility for applicants;

-    Gaining the confidence of applicants;

-    non-discrimination against applicants;

-    observance of confidentiality of information obtained during the certification work;

-    promoting the expansion of the scope of recognition of certificates issued by the Product Certification Body;

-    ensuring satisfaction of consumer requirements.

4.3 Basic principles and general rules of certification

Certification in construction is carried out to protect the interests of the consumer in matters of safety of construction products for life, health, property and the environment, ensuring the reliability of construction structures, increasing the competitiveness of products.

The objects of certification in construction are:

-    project products;

- construction objects - buildings and structures (hereinafter referred to as construction products);

- products of construction and construction materials industry enterprises (hereinafter - industrial products);

- imported products;

- works and services in construction.

Certification is based on the following principles:

- voluntariness;

- non-discriminatory access to participation in certification processes;

- objectivity of evaluations;

- reproducibility of the results of the assessments;

- confidentiality;

- informativeness;

- specialization of certification bodies;

- compulsory verification of fulfillment of requirements for products (services) in the legally regulated sphere;

- reliability of the applicant's evidence of the quality system's compliance with regulatory requirements.

The principle of voluntariness is based on the provision that certification is carried out only at the initiative of the applicant and with a written application, unless otherwise provided for by legislative acts.

All organizations that have applied for certification and recognize the established principles, requirements and rules are eligible for certification. In addition, any discrimination against the applicant and any participant in the certification process is excluded, whether it is excessive price in comparison with other applicants, unjustified delay in terms of time, unreasonable refusal to accept the application, etc.

Objectivity of assessments is provided, firstly, by the independence of the certification body and the experts engaged by it from applicants or other organizations interested in the results of assessment and certification; secondly, by the completeness of the composition of the commission of experts; thirdly, by the competence of experts conducting certification, certified in the prescribed manner.

To ensure reproducibility of evaluation results, verification rules and procedures based on uniform requirements are applied, evaluation is based on evidence, and evaluation results are documented and stored.

Procedures, rules, tests and other activities that can be considered as components of the certification process itself may be different depending on the characteristics of the certification object, which in turn determines the choice of test method, etc. In other words, conformity assessment is performed according to one or another certification system. In accordance with international standards ISO/IEC is a system that carries out certification according to its own rules concerning both procedure and management [20].

Certification of industrial products in construction is carried out on a voluntary basis, except for those cases where the current legislation establishes mandatory certification (windows of various materials, double-glazed windows and locks).

The nomenclature of construction goods subject to mandatory certification is formed by Gosstroy for inclusion in the nomenclature approved by Gosstandart RK.

At carrying out certification, the authorized bodies organize carrying out in accredited testing laboratories (centers) of initial tests of production and assessment of its conformity to the established requirements, and sometimes also assessment and control of the state of production by its certification or certification of the quality system.

Based on the results of testing and assessment of the state of production is issued certificates of conformity and registration of certified products at the presentation of Gosstroy RK in the State Register of certification GOST RK.

Certified products, production or quality systems are followed up by inspection.

Inspection control, as a rule, is carried out by certification bodies that have performed this work, or the relevant territorial supervisory authorities under contracts with the specified product certification bodies.

Inspection control over accredited certification bodies and testing laboratories (centers) is carried out by Gosstroy and Gosstandart RK.

At carrying out certification the conformity of products to the requirements established in the state standards and technical specifications (TS) for products, including the scope of its application (conformity to the purpose), as well as given in building codes and regulations (SNiP) calculation and other characteristics is assessed.

Certification participants are responsible for fulfilling their obligations:

- manufacturer - for ensuring that the products issued by him with a certificate of conformity to the requirements of the relevant ND, for the correct use of the conformity mark and compliance with other conditions of certification;

- testing laboratory - for compliance of its certification tests with the requirements of ND, for reliability and objectivity of the results of these tests;

- certification body - for completeness and correctness of assessment of conformity of products to the established requirements of ND when issuing a certificate and subsequent confirmation of its validity.

4.4 Organization of the state certification system

The organizational structure of the state certification system is formed by:

- authorized body on standardization, metrology and certification;

- accredited bodies for certification of products, processes, works, services;

- accredited testing laboratories (centers);

- accredited organizations providing consulting services in the field of accreditation;

- experts - certification auditors.

The state system of certification provides carrying out of the uniform policy in the field of certification and establishes the basic rules and procedures of certification, requirements to bodies on certification, testing laboratories (centers) and procedures of their accreditation, procedures of preparation and attestation of experts - auditors on certification, rules of maintenance of the register of the state system of certification, audit and inspection control, other requirements necessary for realization of the purposes of certification.

Management of works on certification shall be carried out by the authorized body on standardization, metrology and certification, determined by the Government of the Republic of Kazakhstan.

Physical and legal persons shall organize and carry out their certification activities in accordance with the provisions of this Law and other normative legal acts of the Republic of Kazakhstan, as well as the requirements of normative documents operating in the state certification system.

Accredited bodies on certification of products, processes, works, services and testing laboratories (centers) have no right to render consulting services in the field of accreditation.

4.4.1 Rights and obligations of the authorized body on standardization, metrology and certification

Responsibilities of the authorized body on standardization, metrology and certification:

- creation of the state certification system and organization of its functioning;

- certification of expert auditors;

- maintenance of the register of the state certification system;

- participation in the formation and implementation of a unified state policy in the field of certification, development of perspective plans for the development of certification;

- preparation of lists of products (works, goods and services) subject to mandatory certification;

- accreditation in the state system of certification of organizations for the right to carry out works on certification, certification tests, consulting services in the field of accreditation;

- organization and implementation of state control of activities of testing laboratories (centers) and bodies for certification of products, processes, works, services;

- establishing rules for the recognition of foreign certificates of conformity, conformity marks and test results;

- organization of works on interlaboratory comparative tests (comparisons).

4.4.2 Types of certification of products, processes, works, services

Certification may be mandatory or voluntary.

Mandatory and voluntary certification is established.

Mandatory certification - certification of products, works, services included in the list of products, works, services subject to mandatory certification for compliance with mandatory requirements of a standard or other regulatory document ensuring their safety for life, human health, property of citizens and the environment.

Voluntary certification is carried out at the initiative of applicants (manufacturers, sellers, performers) in order to confirm the conformity of products, processes, works, services to the requirements of normative documents determined by the applicant. Voluntary certification does not replace mandatory certification.

Confidentiality of all information about the organization at all stages of certification and on its results, characterizing the state of the quality system (production) and compliance of personnel, is provided by the management of the certification body both in terms of staff and personnel involved in works on certification.

Specialization of bodies for certification of quality systems (production) is achieved both by the scope of accreditation of the body and by the presence in its staff or among the personnel involved of experts and consultants specialized in the relevant field of activity.

Participation in certification systems can take three forms:

- admission to the certification system;

- participation in the certification system;

- membership in a certification system.

Admission to a certification system means the ability of an applicant to certify in accordance with the rules of that system. Participation and membership in a certification system are established at the level of the certification body. A participant in a certification system is a certification body that applies the rules of the system in its activities, but is not authorized to participate in the management of the system.

Procedures, rules, tests and other activities that can be considered as components of the certification process itself may vary depending on the characteristics of the certification object, which, in turn, determines the choice of test method, etc.

4.4.3 Requirements ensuring product safety

1. Realization of products, works, services subject to mandatory certification without certification of conformity (copy of certificate of conformity) and (or) conformity mark or declaration of conformity (copy of declaration of conformity) shall be prohibited. It is prohibited to advertise products, works, services subject to mandatory certification that have not been certified in the Republic of Kazakhstan.

2. Contracts (agreements) concluded for the supply of imported products subject to mandatory certification shall include

provide for the presence of a certificate (copy of a certificate) and (or) a conformity mark recognized in the state certification system.

Contracts (agreements) concluded for the supply of imported products subject to mandatory certification and intended for retail trade shall provide for the products to be accompanied by information in the state and Russian languages. The information should include the name of the product, country and enterprise - manufacturer, date of manufacture, expiration date, storage conditions, method of use.

4.5 Procedure for product certification

Certification of products in construction is carried out to assess their compliance with the requirements of ND, as well as for compliance with the requirements of international and national standards of foreign countries (the latter - if necessary).

Product certification is carried out by product certification bodies.

Product certification in the construction industry includes:

- submission by the applicant of a declaration-application for product certification;

- consideration of the declaration-application and making a decision on the possibility of certification, including the choice of certification scheme;

- testing laboratory definition;

- drawing up a program and methodology for the certification of these products;

- selection, identification of samples (specimens);

- testing (examination) of products for certification purposes;

- analyzing the status (verification) of production of products;

- analysis of the obtained test results, production verification and making a decision on the possibility of issuing a certificate of conformity and a license to use the Mark of Conformity;

- certification of the production of the certified industrial products or certification of the applicant's quality system (if provided for by the certification scheme);

- execution, registration of the certificate of conformity of production or certificate of conformity of quality system and entering the certificate of production or quality system into the State Register of GOST RK Certification System;

- issuance to the applicant of a certificate of conformity for the production of certified products or for the quality system;

- execution, registration of the certificate of conformity for the products and entering the certified products into the State Register of GOST RK certification system;

- issuance to the applicant of a certificate of conformity and a license for the right to apply the Conformity Mark (or marking of products with the Conformity Mark);

- inspection control over the stability of certified product characteristics, certified production, and quality system;

- information on certified products, certified production and quality systems.

- **.5.1 Rules for product certification**

The procedure of product certification includes several consecutive stages.

Submitting and reviewing a declaration-application for construction certification.

The applicant shall send the declaration - application to the relevant certification body or to the Central Certification Body. The list of accredited certification bodies is published in the Bulletin of Construction Technology (BST). This body registers the application, considers it and not later than 15 days after its receipt informs the applicant of the decision containing the basic conditions of certification, including which test laboratory will conduct product testing, certification scheme, terms of work, and prepares a contract for this work.

The applicant after familiarization with the conditions of certification and agreement with them shall sign the contract. All costs of product certification, regardless of its results, shall be paid by the applicant.

Drawing up of the program and methodology of certification. The program and methodology of certification shall be developed and approved by the certification body, performing this work, and agreed with the applicant. The program and methodology shall take into account the specificity of products and peculiarities of their production and include stages of certification work, the order and rules of their performance, including the rules of decision-making on the results of product testing and analysis of the state of production, terms of performance of individual stages, as well as performers of this work.

Carrying out tests (examination) of products.

Tests of products shall be conducted in testing laboratories accredited in accordance with the requirements of GOST RK. If the testing laboratory is accredited only for technical competence, the tests shall be conducted in the presence of a representative of the certification body or a person authorized by it.

Tests of building structures due to the complexity of their delivery to an accredited testing laboratory may be carried out in the presence of a representative of the laboratory in non-accredited organizations, including the manufacturer, but there must be confidence in the correctness of testing.

Tests of products for certification purposes shall be carried out on standard samples, the composition, design and manufacturing technology of which shall be the same as for products supplied to the customer.

According to the results of product testing, the testing laboratory shall draw up a test report (act of examination), which shall be sent to the product certification body, and a copy to the applicant.

Analysis of the state of production

It is carried out to determine whether the necessary and sufficient conditions exist to ensure the stability of the product in terms of quality. This work is carried out by accredited product certification bodies. They also establish the procedure and methodology for analyzing the state of production. The analysis of the state of production should include the identification of factors affecting the certified characteristics and their stability, including the determination:

- compliance of technical and technological documentation for products and methods of their testing with the requirements of ND;

- quality and sufficiency of control during production, including metrological support;

- the state of technological and production operations that determine the level of certified characteristics;

-   stability of conformity of manufactured products to the requirements of ND;

-   distribution of personnel responsibility for product quality assurance;

-   the state of maintenance and repair of technological equipment.

Based on the results of the analysis of the state of production, a report on the stability of production and product quality at the enterprise is prepared.

Issuance of a certificate of conformity and a license for the right to apply the Conformity Mark of the Certification System

Certificate of conformity for products shall be issued by the certification body for a standard sample of serially produced products, a batch of products or for each specific product on the basis of consideration of the product test report.

The term of validity of the certificate of conformity and the license for the right to apply the Mark of Conformity shall be established by the certification body taking into account the results of tests, analysis of the state of production and the validity period of the ND for the product. In case of changes in design, product formulation or technology, the applicant shall notify the certification body, which shall make a decision on the need for new tests or assessment of the state of production.

When new norms for a certified product parameter are introduced in the ND, the certification body shall decide whether it is necessary to carry out new tests or to assess the state of production of that product. If the manufacturer has not taken appropriate measures to confirm conformity of products to new requirements, the certification body shall terminate the certificate and license. Documents and materials confirming the certification of products shall be kept in the certification body.

Inspection control

Inspection control over stability of certified parameters of output products in the process of its production is carried out by the certification body, which conducted earlier these works. The scope, content and procedure of control shall be established in the Procedure for certification of specific products. Thus periodicity and volume of tests of production shall be established taking into account the results of statistical control of quality of output products.

According to the results of inspection control the certification body may suspend or cancel the certificate of conformity in cases:

\-   violations of the requirements of regulatory documents to which the products are certified;

\-   changes to product NDs or test methods;

\-   changes in product design, composition, raw materials and components used;

\-   changes in the organization and technology of product production;

\-   non-compliance with the requirements of production technology, methods of control and testing, quality assurance system.

The decision to suspend the certificate of conformity shall be taken if, by corrective measures agreed with the certification body, the applicant undertakes to eliminate the identified deficiencies and to confirm their elimination by repeated testing. Otherwise the certificate of conformity shall be canceled. Information on suspension (restoration) of validity of the certificate of conformity shall be communicated by the certification body to the Central Body, the applicant, consumers of products and all participants of the certification system of this product.

4.6 Requirements for certification bodies in construction and the procedure for their accreditation

4.6.1 General Provisions

The body on certification of products, works, productions, etc. can be an organization of any form of ownership, having the status of a legal entity, not having certain power (control) functions, not dependent on manufacturers and consumers of products and not having administrative or other influence on the results of certification activities, having the necessary competence in the field of development, manufacture and certification of these products, works, services, productions, quality systems in construction.

An organization seeking accreditation and operation as a product certification body shall have the necessary facilities and documented procedures in place, including:

- qualified and specially trained personnel;

- a fund of regulatory documents establishing the procedure for inspection control of certified products, certified production;

- register of certified products, works, services, productions, quality systems;

- organizational and economic opportunities (conditions) for carrying out certification works.

Work on certification is carried out at the request of organizations and enterprises manufacturing or supplying the given product (service, etc.). The scope of accreditation of the certification body includes the nomenclature of certified objects, characteristics (parameters) confirmed during certification, the list of normative documents, for compliance with which certification is carried out, and the list of normative documents for test methods.

The head of the certification body and specialists, carrying out management of works on the main activities of the body, as a rule, should be certified expert-auditors of the GOST RK Certification System.

4.6.2 Functions of the certification body

The main function of the certification body is to carry out certification of products, services, productions, quality systems of certification objects in accordance with the scope of its accreditation, including:

- receiving and reviewing applications for certification;

- selection of the certification scheme, determination of the certification procedure;

- organization and carrying out of certification of certification objects;

- execution of certificates of conformity, their accounting and transfer to the Central Body for Certification in the field of construction for entering into the State Register of GOST RK Certification System;

- issuance to applicants of a certificate of conformity and a license to apply a certificate of conformity;

- organizing and conducting inspection control of certified products, certified production, certified quality systems;

- keeping records of certified products and preparing publications on certification results;

- revocation or suspension of the certificates of conformity issued to them.

The functions of the certification body are also:

- formation of a fund of NDs used in certification, their timely updating;

- development and maintenance of organizational and methodological documents establishing the rules of functioning of the certification body;

- collection, storage, analysis and systematization of information on the quality level of certified domestic and foreign certification objects;

- interaction with applicants on timely adjustment of the quality of certified certification objects in case of changes in the requirements of ND.

4.7 Requirements for testing laboratories in construction and the procedure for their accreditation

4.7.1 General Provisions

The organization applying for accreditation and functioning in the GOST RK Certification System as a testing laboratory (center) shall be competent, independent from manufacturers and consumers of the tested products, have the necessary facilities and documented procedures to conduct product testing for certification purposes, including:

- test equipment and measuring instruments, facilities, etc.;

- qualified and specially trained personnel;

- legal, organizational and methodological documents establishing the procedure and rules for conducting product testing and ensuring the quality of testing;

- fund of normative documents for products and methods of their testing.

Accreditation of testing laboratories in the GOST RK Certification System is an official recognition of their technical competence in testing specific types of products, as well as the independence of the laboratory from developers, manufacturers (suppliers) and consumers of products.

Testing laboratories conduct product testing on behalf of certification bodies, based on decisions taken by the latter on declarations-applications for product certification.

The scope of accreditation of the testing laboratory includes the nomenclature of tested products, nomenclature of determined parameters.

The testing laboratory shall be responsible for the objectivity of testing and reliability of the results obtained, for ensuring the confidentiality of information obtained as a result of product testing.

Requirements to the personnel of the testing laboratory, to the premises for testing, testing equipment and measuring instruments, their certification and verification, made during testing, shall comply with the requirements of the GOST RK Certification System, established in the documents of the System. The use by the testing laboratory of certified testing equipment and verified measuring instruments of other organizations is allowed in exceptional cases, for example, in case of temporary malfunction of its equipment and only by agreement with the certification body, based on the decision of which the tests are conducted.

4.7.2 Legal status of a testing laboratory

The legal status of the testing laboratory shall provide it with the rights of an independent legal entity and independence of actions in the field of testing, excluding administrative, financial and commercial dependence on manufacturers and consumers of products.

If the testing laboratory is not a legal entity, but is part of an organization (certification body) that is a legal entity, it shall be a structural subdivision of this organization (certification body), provided that the organization is independent of manufacturers and consumers of products.

The testing laboratory, conducting tests of products for certification purposes, can not be a structural subdivision of the organization-manufacturer or consumer of products. The testing laboratory shall be staffed with a permanent staff of specialists having appropriate education, professional training, including special training, qualification and experience in testing and quality control.

4.7.3 Functions, rights, duties and responsibilities of the testing laboratory personnel

Functions, rights, duties and responsibilities of the personnel of the testing laboratory, requirements for technical knowledge and work experience shall be established by job descriptions or other internal documents of the laboratory, which shall be reviewed in a timely manner.

The structure of the testing laboratory may provide for the presence of independent subdivisions conducting tests of separate groups of products or separate types of tests.

The testing laboratory may be part of the product certification body as its subdivision.

The head of the testing laboratory and specialists, who supervise the work on the main activities of the laboratory, shall be certified expert-auditors of the GOST RK Certification System.

The main function of the testing laboratory is to conduct tests of construction products for certification purposes according to a fixed nomenclature and types of tests.

The test laboratory in accordance with the decision of the certification body may also conduct sampling of products for testing, analyze the state of production of certified products, participate in the certification of production and quality systems of manufacturers and inspection control of certified products, production, quality systems.

4.7.4 Requirements to the documents of the testing laboratory

Accredited testing laboratory shall have a set of legal and organizational-methodological documents, ensuring its functioning in the GOST RK Certification System during product testing. The set of documents shall include:

- regulations on the testing laboratory;

- passport of the testing laboratory;

- accreditation certificate with the scope of accreditation;

- the quality manual of the testing laboratory;

- documentation of test equipment and measuring instruments;

- ND regulating the requirements to the tested products and test methods;

- job descriptions of the personnel of the testing laboratory.

The testing laboratory shall have:

- a recording and logging system that should show how each test was conducted;

- **a** system for receiving and reviewing the certification body's test decisions or product test applications and issuing the results thereof;

- procedure for collecting, storing, analyzing and systematizing information on the quality level of domestic and foreign products.

4.7.5 Procedure for accreditation of a testing laboratory

Organizations can be accredited as a testing laboratory: research institutes, design and development bureaus, joint-stock companies, as well as if they meet the requirements of independence from manufacturers and consumers of products.

Organizations belonging to associations which also include manufacturers or consumers of products and which are not third parties in relation to the applicant may not be accredited as a testing laboratory.

Procedure for submission and review of applications for accreditation of a testing laboratory in construction

Any organization meeting the requirements set out above may apply to the Head of the Central Certification Body for Construction for accreditation as a testing laboratory.

The application shall be accompanied by:

- Draft Regulations on the Testing Laboratory;

- passport of the testing laboratory;

- the quality manual of the testing laboratory;

- procedure for preparation and conduct of product testing for certification purposes in a testing laboratory;

- a copy of the Charter of the applicant organization or its registration certificate;

- order on organization of the testing laboratory and its preparation for accreditation.

The draft Statute of the Testing Laboratory shall contain the following sections:

- area of accreditation;

- legal status, administrative and organizational structure of the testing laboratory;

- functions of a testing laboratory;

- rights, duties and responsibilities of the testing laboratory;

- interaction of the testing laboratory with other organizations;

- information on expert auditors.

Procedure for reviewing documents required for certification

The Central Certification Body shall consider the application and conduct an examination of the submitted documents. In case of positive results of the examination the Central Certification Body shall prepare an order on appointment of the commission on verification of conformity of the testing laboratory to the requirements of the GOST RK Certification System and work program of the commission on verification of the testing laboratory, and in case of negative results - a decision with a motivated refusal.

Verification of compliance of the testing laboratory with the requirements of the GOST RK certification system

The inspection shall be carried out directly at the accredited testing laboratory in accordance with the inspection program.

In the process of verification of the testing laboratory shall be carried out control tests of products in accordance with the scope of accreditation. According to the results of the Commission's work the act of verification of conformity of the testing laboratory to the requirements of the GOST RK Certification System shall be drawn up, which reflects the state of affairs on all issues of the verification program and gives recommendations on the possibility of accreditation or refusal thereof.

Procedure for preparation of documents of the testing laboratory for coordination and approval, execution, registration and issuance of accreditation certificate

On the basis of the Act of the Commission, the Central Certification Body or on its behalf the Center for Certification in Construction prepares for approval the Regulations on the testing laboratory, draws up the accreditation certificate and accreditation area, draws up the license agreement, which are submitted for approval to Gosstroy RK. After approval of these documents the testing laboratory is registered in the State Register of GOST RK Certification System. Registered accreditation certificate and approved documents of the testing laboratory, as well as the license agreement shall be sent to the applicant organization.

5   List of sources used

1. Zubkov V.A., Sviridov V.N., Nagornyak I.N., Treskina G.E. Standardization and technical standardization, certification and testing of products in construction : textbook. - Moscow: ASV Publishing House, 2003. - 224 c.

2. Krylova G.D. Fundamentals of standardization, certification, metrology: textbook for universities. - 2nd ed., revision and additions. - M.: UNITI-DANA, 2001.-711 p.

3. Lifits I.M. Fundamentals of standardization, metrology and certification : textbook. - 2nd edition, revised and supplemented. - M.: Yurait-M, 2001. - 268 c.

4. Nagornyak I.N., Belan V.I., Soshkina G.I., Sviridov V.N. et al. GOST R certification system. Certification of products in construction: Manual. - M.: GUP TSPP, 2000. - 189 c.

5. Okrepilov V.V. Quality management : textbook for universities. 2nd ed., supplement. and revision. - SPb: Nauka, - 2000. - 912 c.

6. Law of RK "On standardization" from June 16, 1999 with amendments from June 10, 2003 Law of RK "On ensuring the uniformity of measurements"; Law of RK "On certification" from June 16, 1999 with amendments from January 15, 2001, from July 11, 2001, from December 15, 2001, from June 10, 2003.

7. ST RK RK 1.0-2000 "GSS RK. Basic Provisions", ST RK 1.1-2000 "GSS RK. Standardization and related activities. Terms and definitions", ST RK 1.2 -2002 "GSS RK. The order of development of state standards".

8. Ovsyannikov O.A., Razu M.L. Organization of management in construction. M: V. shk., 2007.

9. Fedko BJL, Albekov A.U. Marking and certification of goods and services: Manual, Rostov-n/D: Phoenix, 2008.

10. ST RK ISO 9000 -2001 "Quality management system. Basic provisions and vocabulary".

11. ST RK ISO 9001-2001 "Quality Management System;.

Requirements."

12.	CTRK	1.0-2000	"	State System of Standardization

.

Basic Provisions."

13.	STRK	1.4-1999	"	State system of standardization

of the RK .

Firm Standards. Basic Provisions."

14. ST RK 1.12-2000 "State Standardization System of the Republic of Kazakhstan. Normative text documents".

15.	STRC 2.1-2000	"	State system of standardization

of the KR .

Terms and Definitions."

16.	STRC 2.3-1997	"	State system of standardization

of the KR .

Standards of units of physical quantities. Basic Provisions".

17. ST RK 2.18-2001 "State Measurement System of the Republic of Kazakhstan. Methods of measurement performance. Procedure of development, certification and application".

18. GOST 8.010 "State system of measurements. Methods of performing measurements".

19. GOST 8.401-80 "GSI. Accuracy class of measuring instruments. General provisions".

20. GOST 8.417-81 "GSI. Units of physical quantities".

I want morebooks!

Buy your books fast and straightforward online - at one of world's fastest growing online book stores! Environmentally sound due to Print-on-Demand technologies.

Buy your books online at
www.morebooks.shop

Kaufen Sie Ihre Bücher schnell und unkompliziert online – auf einer der am schnellsten wachsenden Buchhandelsplattformen weltweit! Dank Print-On-Demand umwelt- und ressourcenschonend produziert.

Bücher schneller online kaufen
www.morebooks.shop